William Manning

Recollections of Robert Houdin

Clockmaker, Electrician, Conjuror

William Manning

Recollections of Robert Houdin
Clockmaker, Electrician, Conjuror

ISBN/EAN: 9783741170355

Manufactured in Europe, USA, Canada, Australia, Japa

Cover: Foto ©Andreas Hilbeck / pixelio.de

Manufactured and distributed by brebook publishing software
(www.brebook.com)

William Manning

Recollections of Robert Houdin

To Our Professional Friends.

It may interest many of you to know how we gained possession of the little book entitled "Recollections of Robert-Houdin, by William Manning."

A cousin of ours, Mr. Ralph Meriman, a well-known artist, resides in Paris, where he has a very extensive acquaintance. We requested him to call on the widow of Emil Robert-Houdin and obtain particulars from her about this little brochure. Mr. Meriman succeeded in procuring a copy of the book, through the kindness of Mr. Manning, the author, and as the means by which it all came about are very interesting, we publish herewith the entire correspondence relating thereto.

We are under special obligations to Mr. William Manning, for his kind permission to republish his little work, and we have no doubt it will prove of great interest to the conjuring fraternity at large. We republish herewith the entire book excepting the part devoted to the minute explanation of Houdin's Electrical Clock, which would not be of much interest to our readers now, there being so many things in that line that are more modern in these days of rapid improvements in electricity. The following is a copy of a letter to Mr. Ralph Meriman from Madame Emile Robert-Houdin:

Paris, Jeudi, 21 Avril, 1898.

Monsieur Meriman—Selon votre désir je vous envoie une petite brochure faite par un ami, et qui vous donnera tous le renseignements que vous desirez avoir sur "Robert-Houdin."

Veuillez en avoir le plus grand *soir*, et me *le retourner le plus tot possible* après en avoir pris connaissance. Vous trouverez les oeuvres de Robert-Houdin, 18 Boulevard des Italiens à la Libraire Nouvelle.

1

Veuillez agréer Monsieur, mes salutations les plus distin-
gué.
(Signed)   Ve. Emile Robert-Houdin.

TRANSLATION OF ABOVE LETTER.

Paris, April 21, 1898.

Mr. Meriman:—According to your wish I send you a
little book, written by a friend, and which will give you all
the information you desire.

Please be very careful with it and return it as soon as
possible after noting its contents.

You will find the works of Robert Houdin at 18 Boulevard des Italiens in the New Library.

Accept dear sir, my best regards,
                    Yours,
                    Ve Emile Robert-Houdin.

———

First letter to Mr. Ralph Meriman from Mr. W. Manning:

Worthing, April 30, 1898.

Dear Sir—Your letter was wrongly addressed to me.

I fear you will have difficulty in procuring a copy of my
booklet, as only a limited number were printed. I saw one
in a bookseller's list lately for 21 shillings. Would you like
me to try and procure one? Being out of print it has fetched
as much as 42 shillings, and unfortunately by the rules
of my Club, no reprint is permissible, or there would have
been a large sale; and no profit has ever been made by the
publication except what the second-hand bookseller may
make.

I am much pleased that you liked the brochure and regret
that I cannot myself supply your wants.
                    Yours faithfully,
(Signed)   W. Manning.

———

Copy of postal card to Mr. Ralph Meriman, Paris:

London, May 13, 1898.

The little book is unique and the illustrations would be
difficult to reproduce. The work is privately printed by the

Sette of Odd Volumes and is not copyright. Extracts, or the whole discourse, with an acknowledgement of the source of origin would not be objected to.

(Signed)   W. MANNING.

Copy of postal card to Mr. Ralph Meriman:

LONDON, June 7, '98.

I have thought it best to secure the book.

Yours,

(Signed)   W. MANNING.

Copy of last letter from Mr. W. Manning to Mr. Ralph Meriman in Paris:

LONDON, June 9, 1898.

DEAR SIR—Our letters seemed to have crossed and you had not, when writing yesterday, received my card saying that I thought it best to buy the book before hearing from you.

The difficulty with such a limited issue is to find one at all, and I told you I could have procured one at 42 shillings, but I tried and succeeded in doing better for you.

Only today I was asked *three guineas* for one of these *opuscula*, only two years older than my own! It has not half the matter, has no illustrations, and as a smaller number were issued than of my own, the price has risen, for book collecters are not satisfied without getting a *perfect sette* of all the issues.

I shall greatly value the book you speak of, and in the future *to see what use has been made* of my modest little volume! If you please.

I have sent the book to City as you suggested, and I am glad, indeed, to have been of any service to a friend of Madame Veuve Emile Robert-Houdin.

Yours sincerely,

(Signed)   W. MANNING.

Copy of letter to Messrs. Longsdorf, Meriman & Co.:

LONDON, June 9, 1898.

GENTLEMEN—Mr. Meriman, of Paris, has asked me to

forward the accompanying booklet to you, and says he has instructed you to pay the cost incurred, 28 shillings.

Yours faithfully,

(Signed)   W. MANNING.
(Received 28 shillings this date.)

Copy of letter to Mr. H. J. Burlingame, Chicago:

LONDON, July 22, 1898.

DEAR SIR—You have given me much pleasure in sending me so large a parcel of conjuring literature, for I still retain my old interest in the art. It may interest you to know that at one period of my life the great master—Robert-Houdin—finding me an apt imitator and inventor, proposed to my guardian to take me as a pupil, and adopt me as his successor!

I shall hope to get a sight of my little brochure, the "Recollections," in its new form, if you incorporate it in any future work.

Believe me yours sincerely,

(Signed)   W. MANNING.

The little book is 4½ by 5½ inches in size, printed on hand laid paper with deckel edges, and contains 81 pages. The last 14 pages consist of a biography of the privately printed Opuscula, issued to the members of the Sette of Odd Volumes, a list of members of "Ye Sette of Odd Volumnes," and of "Supplemental Odd Volumnes."

On the outside of front cover is the visiting card of

as written by himself. On the outside of back cover is Robert-Houdin's Seal.

4

ROBERT-HOUDIN'S SEAL.

The first printed page inside is as follows:

"Privately Printed Opuscula

Issued to the Members of the Sette of Odd Volumes

No. XXIV.

RECOLLECTIONS

OF

ROBERT-HOUDIN.

Clockmaker—Electrician—Conjuror."

Then on one of the following pages is a magnificent, though small portrait of Robert-Houdin, a reproduction of which we show on next page.

ROBERT-HOUDIN.

Opposite this illustration is the title-page, as follows:

"RECOLLECTIONS
OF
ROBERT-HOUDIN.
BY
WILLIAM MANNING,
Seer
To the Sette of Odd Volumes.
Delivered at a Meeting of the Sette held at Limmer's
Hotel, on Friday, December 7,
1890."

"Imprinted at
THE CHISWICK PRESS, TOOKS COURT,
Chancery Lane,
London.
MDCCCXCI."

On the second following page is the dedication:

"DEDICATED

TO THE PRESIDENT

AND THE SETTE OF ODD VOLUMES."

On the second next page appears the following:

"THIS EDITION IS LIMITED TO 205 COPIES AND IS
IMPRINTED FOR PRIVATE CIRCULATION ONLY.
No. 103.
Presented Unto

By
W. MANNING."

There being a blank for the receiver's name and the auto graph of Mr. W. Manning.

On the second next page appears:

On the second next page appears the introduction proper, as follows:

"TO THE EVER COURTEOUS READER.

"I have to thank the President and brethren for inviting me to print this address, and I regard it as a great honour that my own little volume will take its place among the more famous ones which have preceded it.

"I owe especial thanks to our distinguished *art critic* (Brother G. C. Haité) for suggesting that my son, W. W. Manning, should execute some illustrations for this *opusculum*, and I shall be deemed wanting in personal regard if I omit to thank him also for the willing help he has given me in placing before our brethren and guests some real recollections of the December meeting.

"WILLIAM MANNING,
"Seer to the Sette of Odd Volumes."

February 6th, 1891.

On next page appears

"ROBERT-HOUDIN,

Born at Blois, Dec. 6, 1805.

Died at Blois, June 13, 1871."

---

*Reproduced from the original sketch, kindly lent by Mdme. Veuve Emile Robert-Houdin.

Then commences the body of the work:

"RECOLLECTIONS OF
ROBERT-HOUDIN.

*Your Oddship, my Brethren of the Sette of Odd Volumes, and
Guests:—*

When I promised the President in obedience to his invitation to read a paper, that I would give some Recollections of Robert-Houdin, I had no idea that I had materials sufficient for a biography.

And when I thought of illustrating my reminiscences with a few experiments, most of which are personal memorials of the arch-conjuror, I feared that exhibition of old tricks might fall flat, or be deemed an impertinence.

I hope to surmount these two difficulties by condensing my remarks into the smallest possible space, and by craving the indulgence of Brethren and Guests for producing any experiment with which they may be familiar.

I further ask for your generous forbearance if I appear to speak unduly of myself, for I do assure you that my only aim to-night is to glorify my hero.

Well then, to begin quite at the beginning, I made the acquaintance of Robert-Houdin and his interesting family when I was a school boy.

During his stay in London in 1849, I was an almost daily visitor at his house, and my intimacy and correspondence with him continued to nearly the last month of his life.

In those early days I was the playfellow of his two sons, Emile and Eugène, and sometimes assisted the family in making up the freshly-cut flowers from Covent Garden into the small button-holes, which were to play their part in the evening's performance. I really must introduce you to the family circle when I first entered the magic ring at 35, Bury Street, St. James's.

The ring consisted of

Monsieur,
Madame,
Emile,
Eugène.

These comfortable apartments had been secured by my

valued friend, the late John Mitchell, whose keen sense of business had led him to make handsome overtures to the Parisian prestidigitateur, and as the French plays were then running at St. James's Theatre three nights in the week, to fill up the other three nights with the Soirées Fantastiques of Robert-Houdin.

I remember the first occasion of my dining with the family, that Mdme. Robert-Houdin, whose knowledge of the English language was extremely limited, but whose solicitude for her guest was unbounded, made an attempt to draw me out upon the subject of a dish, which she hoped might be congenial to my taste. At that time she was mistress of about four sentences in English, which she repeated with the precision (I speak with all respect) of the parrot. To the intense amusement of her husband she looked inquiringly in my face, and said with great deliberation and excellent pronunciation "I love you!" She was a very pretty woman, and I appreciated that mark of her favor; but I was a very small boy, and scarcely understood the merriment of Monsieur, when roaring with laughter, he explained to me that she didn't really love me, but had employed one of her four sentences, and that what she really meant to ask was, "Do you like it?" No ill feeling arose from this mistake, and as I was not only small but unsophisticated, my visits to this happy family were not interrupted by the incident.

It was on this occasion that I saw the first drawing-room trick, as played by the master-hand, and if space allowed, I would exhibit it to you. He would take a handkerchief thus, and thus, making knots and failing to make them, to the bewilderment of the spectator, and though as you perceive it is still perfectly empty, I take it up, and by simply shaking it thus, I fortunately am enabled to produce this shower of goodies, which I have much pleasure in sending around for the gratification of the Odd Councillors.

If your Oddship, always so particular on the subject of language, but always so forgiving to an erring brother, would permit me to mutilate a famous proverb, and blend three tongues into one, I would say of this shower of sweet-meats:

"DE MORTUIS NIL Nicey *bon bons!*"

With regard to the sons, Emile and Eugene, I was only too ready to be the friend and *confrère* of two individuals, who at that time were playing their respective parts in exciting the wonder and admiration of the London public at St. James' Theater.

Emile, the older of the two sons, assisted his father on the stage with a manner and quickness peculiarly French. He took part in the wonderful vanishing trick, in which he was placed upon a table on the stage, and being covered with a huge extinguisher, his father, on firing a pistol at the table, overthrew the extinguisher; while in a few seconds another pistol-shot from an unoccupied box at the back of the theater attracted all eyes, when was seen the form of the boy Emile, bowing to the audience.

Emile also took part in the performance of the Inexhaustible Bottle trick, which at that time produced a great disturbance in the public mind, but which, as you will see by my model, is very simple, its action being pneumatic.

In addition to the bottle itself, from which a hundred glasses of liquor were handed to his clamoring audience, I may tell you confidentially that the glasses were infinitesimally small, and for noyeau or other liquors which are white, many of those glasses were already properly and fully charged, and the filling was sheer make-believe on the part of the conjuror.

This will be a convenient opportunity for saying that in later years, and after his father retired from a profession for which Emile had no real love, the latter became a watchmaker in the house of M. Breguet, where he greatly distinguished himself as a workman of the first rank, and on his marriage to one of the most charming of women, was established as a watch and chronometer maker in Paris, turning out instruments of precision of the very first order. He was afterwards induced to take up the old theater, where his father had made so much fame and so much money, and he held this property until his death, which occurred after a few days' illness in 1883.

He was a man full of good nature and *bonhomie*, and his name will ever be associated with that of his father as his great helpmate in that masterpiece of modern magic with

11

which most of you are familiar from the frequent publication of its details, "Second Sight," in which performance, with bandaged eyes, he gave the audience, with unfailing accuracy, the name and full description of any articles which his father might take from the hand of a visitor.

Emile Robert-Houdin published a treatise on clock and watch-making, to which his father wrote the following preface:

"On m'a souvent demandé pourquoi mon fils au lieu de suivre la carrière que je lui avais ouverte dans la presdigitation, avait préféré se livrer à l'étude de l'horlogerie. Ce que j'ai répondu dans cette circonstance peut avoir un certain à-propos en tête de cette brochure.

Si l'on admet les vocations héréditaires, c'est bien le cas d'en faire ici la juste application: le bisaieul maternel de mons fils, Nicolas Houdin, était, au siècle dernier, un horloger d'u grand mérite. J. F. Houdin, sons fils, a conquis, on le sait, une des premières places parmi les horologers les plus distingués de notre époque. Certaine réserve que l'on comprendra ne me permet pas de faire avec autant d'abandon l'éloge de mon père: je me contenterai de dire que c'était un horloger très adroit et très ingénieux. Quant à moi, avant de me livrer aux travaux prestigieux dont le mécanisme était la base, je me suis occupé, longtemps, d'horlogerie de precision et, dois-je le dire, j'y ai obtenu quelques succès.

Avec une telle généalogie, peut-on ne pas être predestiné a l'horlogerie? Aussi mon fils, entraîné par une vocation irrésistible, s'est-il livré sans réserve a cet art qu'ont illustré les Berthoud, les Breguet; et c'est près du dernier de ces deux célèbres maitres qu'il a connu les éléments de la profession de ses ancêtres.

ROBERT-HOUDIN."

TRANSLATION OF ABOVE.

"I have often been asked why my son did not follow the career I had opened for him in prestidigitation, but preferred instead the study of horology. My answer to the question may be used fitly as a preface to this pamphlet.

If you believe in hereditary vocations, here is a case for their just application. My son's maternal great grandfather,

Nicolas Houdin, was a watchmaker of great merit in the last century. J. F. Houdin, his son, has gained, as is well-known, a prominent place among the most distinguished watchmakers of his time. A certain modesty, which you will understand, prevents me from praising my father as highly; I shall only say that he was a very skillful and ingenious watch-maker. Before devoting myself to the art of conjuring, based on mechanism, I, too, was for a long time a watchmaker and achieved some success.

With such a genealogy, should one not be predestined to horology? Therefore my son was irresistibly drawn to his vocation and he took up the art which Berthoud and Briguet have made famous. It was from the latter of the two celebrated masters that he learned the elements of the profession of his forefathers.

ROBERT-HOUDIN."

Eugene was a younger son, and appeared at St. James' Theatre in the trick known as the "Suspension by Ether," the latter drug being then only recently in vogue as an anesthetic. Houdin led his handsome boy by the hand to the footlights to make the most mechanical of bows to his audience. The two slowly retired backwards, when the father fixed an upright rod under each arm of the son, who had ascended three steps for the purpose of raising himself from the stage. The father then expatiated gravely upon the marvels of ether, and pretending to administer it to the youth, a simulated slumber followed, and the steps being suddenly removed, the boy remained supported by the two rods only, his body retaining its vertical position, the feet eighteen inches from the stage. Houdin then very carefully raised the body to the horizontal line without disturbing the slumber of the boy, and to the terror of many a spectator, the father suddenly kicked away rod number two, leaving Eugene's outstretched body apparently without a support, his right elbow only just in contact with rod number one. My illustration represents a further development of the experiment which appeared to defy the laws of nature. This was always the final trick of a performance, and when the curtain fell, and was raised again in obedience to the recall, father and son came walking most gravely forward, and the effect of this slow movement was to make half the world believe that the boy was not flesh and blood at all, but a marvellous automaton!

This same Eugene played an important part in his country's history.

A most interesting letter, the last I received from his father, narrates his death so graphically that I insert it further on in this discourse.

This may be a convenient moment to show you one or two of the smaller souvenirs of my hero, and, as I alluded to flower-cutting, perhaps this box, a well used one at the theatre in the Palais Royal, as also at St. James' Theatre in King street, may be presented for your inspection.

Wait a moment! I had intended that it should have contained natural instead of the artificial flowers with which you see it is crammed. You will pardon the error, for these are worthless as gifts, their perfume being lost. I empty them upon my desk, and close the box tightly thus, and open it instantly thus; and I invite you to observe that it is full to overflowing with real roses and sweetly scented violets, to which you are truly welcome. Stay!

"There's rosemary, that's for remembrance!"

—or rather for the *Remembrancer*, in acknowledgment of his gracious gift of tonight[*], and forget-me-nots for all!

Those among my audience who have read the Memoirs of Robert-Houdin[†] may remember that the accidental purchase of the wrong book handed to him by an oblivious bookseller (the London bookseller is never oblivious), who gave the young boy a treatise on magic, when he was paying for one on botany,—this accident, I say, turned the whole current of his mind in the direction of trick-making, his first effort being devoted to the manufacture of toys in endless variety.

He gave me one which at that time could be purchased from the Paris toy-makers, and which I will endeavor to show to you. It is in good preservation, though it has been in my possession since 1849, religiously taken care of as you

[*] "Neglected Frescoes in Northern Italy," O. V. Opuscula, No. XXIII, by Brother Douglas H. Gordon.
[†] Chapman & Hall, London, 1850.

14

perceive, and its delicate limbs have remained unbroken, for it was exhibited to my children on state occasions only. As I take it from its little box, by a reversal of the lid I build up a platform with three stages, and placing my old friend on the top of the highest platform I gently blow in his face, and as you see he turns somersault after somersault until he reaches my desk, when he obliges us with a final fling expressive of indignation that there are no more steps left! I pass this fantastic acrobat round. It appears to be of eastern descent. Your Oddship will observe that the dress is scanty, but picturesque; but as your experience is vast in these matters, you will, perhaps, determine the nationality, with further particulars for the benefit of inquisitive brethren.

Another of his gifts and of his inventions also lies at my hand—a negro's head, a gentleman from Darkest Africa, which, as you see, will not permit itself to be severed from its body by my knife. How useful such a head would have been to many a notability in our own dear history, for as your Oddship perceives, this head will not come off, cut which way I please.

> No knife can cut this throat in twain,
> No juggler rend this jug'lar vein!

I much regret that I have no illustration for this popular person, but during the month at my disposal the days were as dark as the nights and the backgrounds were too black for his sable skin.

During his stay in London, Robert-Houdin enjoyed a long series of of triumphs, and was the sensation of the season, producing novelty after novelty. He made a tour of the provinces, and visited Ireland and Scotland before his final return to London, and with a daring quite characteristic of his determination to surmount difficulties, he addressed all his country audiences in English, producing many an encouraging cheer and roar of laughter by some of his mistakes in the choice of words.

On his provincial tour, he paid me a visit at my school in Birmingham, and I must here give you an example of his kindness and manner of displaying it.

Never can I forget a certain Sunday during my friend's stay in Birmingham. He was then performing at Dee's Hotel, and had comfortable quarters in the immediate neighborhood. After dinner many friends dropped in, and after a general conversation, it was determined to play at cards, and among other guests, I was invited to sit at the round table.

I had been brought up as a strict observer of the Sabbath, and I took my seat with a sort of holy dread! Scarcely were the cards dealt out, and while I was still in the throes of horror at my new position, when the sound of evening bells from the neighboring church of St. Philip added to my feelings of remorse that I should be engaged in such unholy pastime. "Bells! bells! bells! bells!" This game in which I was engaged and the language of my host and his friends were alike foreign to me.

In addition to the remonstrating clanging of the deeptoned bells, I was also conscious of getting deeper into the mental mire by the adverse run of luck. "Bells! bells!" were ringing in my ears, and "Losing! losing!" was making itself audible in the innermost recesses of my heart. My counters at every moment were getting sensibly fewer, and, in addition to feeling that I had unexpectedly, and without malice prepense, dived into the vortex of dissipation, I was conscious of being unable to meet my engagements when should come the moment for final settlement, as we were playing for money, and I had nothing in my schoolboy pockets but the proverbial knife and piece of string.

The suspense was more awful than I can describe. The end came all too soon. Ill luck pursued me with a continuity that was relentless, and when my despair was at its greatest, and my exchequer at its lowest, the game suddenly terminated, and I was all but bankrupt in counters, and absolutely so in pocket. My host and friend was my vis-a-vis, and probably had observed my distress, for, as if by very magic, with a face full of fun, he took a handful of his winnings, and stretching across the table placed a good pile of pearl fishes before me, which exactly enabled me to pay twenty shillings in the pound, and to rise up from my seat blessed with those finest of human emotions—"Peace with Honour!"

After his return to Paris, a few years enabled him to

amass a handsome fortune, and seek retirement in his native town of Blois. But even there he was ever at work. Among many wonders, he made a watch with a pedometer movement which was always winding itself up, that is to say, when the wearer moved about all day, it received sufficient winding impetus to go on through the night; so long as the owner walked up, the watch would never run down.

In his later years he devoted himself to the subtler sciences, and read many original articles before the learned societies, copies of which I possess. He was a master of the science of optics, and he presented me with an opthalmoscope which he invented for the examination of his own retina.

AUTOMATA.

I have by me a list of the chief of his automata, most of which I have seen, but which it is tantalizing to describe, as I cannot show them. The greater number were made during those early struggles to which so many famous men have been born. (Indeed, it occurs to me that, without them, one may scarcely hope to become great!)

I talk at random, merely to mention them:

The Chinese Juggler.

Auriol and Debureau.

The Mysterious Orange Tree.

A tumbler who performed daring tricks upon the tra peze

The Pastry Cook who distributed cakes and wine to the spectators.

The Writing Automaton.

Just a word about his writing automaton, perhaps the most marvellous of his works, which answered questions proposed by spectators, and even drew elegant emblematic designs in reply to some questions.

The artist had many disappointments in procuring a suitable head for his new automaton, for the sculptor had produced him an admirable model for the body of his figure-but, being a maker of saints, had put rather too much sanctity into the face to give satisfaction. So after many fruitless efforts to obtain what he required, Houdin set to work for himself, and, with a ball of modeling wax and a looking-glass, actually succeeded after many disappointments in making an excellent model of his own good face.

Houdin showed this, his first marvel of mechanism, to one of his servants, and used to tell the tale again and again that the man was highly complimentary, stating that he, too, knew something of mechanics, as he always had to grease the vane on the church steeple.

He chose a residence in the country for the construction of his next great triumph, a nightingale, whose delicate pipings were reproduced with marvellous fidelity, and whose beak opened and closed in time with the notes it was producing, and whose body lept from branch to branch of the trees by which it was surrounded.

My space is so limited or I should have shown you with what dexterity he produced innumerable plumes from a small square of velvet, similar to this that I hold before you. As I have no plumes I cannot produce them. He would throw the velvet thus, across his shoulder, and, instantly withdrawing his hand, would hold aloft a large bowl of gold fish, as, happily, I am able to do now!

Bowl, fish and water, all real.

Look at his MYSTERIOUS CLOCK. Nothing could be simple

pler in appearance, yet few problems are more difficult of solution. The small base is surmounted by glass, transparent. The face is of glass also, and equally transparent with the cylinders which support it. The dial has but one hand. There is no connection but glass between base and dial, and yet the clock is a perfect timekeeper.

Do you not agree with me that the man who conceived this masterpiece of deception must have had an imagination of no common order, and to have executed it with his own hands in a manner which defied detection, must have had brains almost at his fingers' ends? In the preceding paragraph I have given you *one letter* which may be a key to the clock mystery. DO NOT GIVE IT UP!

GHOST ILLUSION, ETC.

In the midst of the pleasures of his retirement he was ever taking the liveliest interest, not only in his own particular inventions, but in the tricks, illusions and deceptions which were being produced in England, and I had many communications with him on these subjects, furnishing him with drawings or models, many of which he reproduced with startling improvements and additions in his grounds at St. Gervais. The Ghost Illusion was a secret well kept by the people at the Polytechnic, as well as the famous Sphinx by Stodare at the Egyptian Hall; but I give you my word that my friend received full particulars at the earliest moment possible, and I carried my regard to this extent, that I not only investigated the so-called supernatural powers of the child known as the Infant Magnet, but, on a public stage, during the performance of certain mysterious phenomena by a young lady who shall be nameless, I consented to be locked up in a dark cabinet with that interesting maiden, whose toilette was superb.

I was searching for Truth in the interests of Science. I was attached to that young person with tapes. I remember on my return home rather late to our apartments at W——, SOMEONE said to me:

"Where have you been?"

I dissembled and said, "N—owhere."

She said, "But what is this white stuff upon your sleeve?"

Again I dissembled and said:

"Oh, it rubs off!"

Said she, "I believe it is powder."

"Gunpowder?" said I.

"*Is gunpowder white?*" said she.   (She had never been sarcastic before.)

So I had to make a clean breast of it, and told her exactly what I have told you.

I never betray Cabinet secrets, but my report to my correspondent was that the phenomena which had taken place in the dark cabinet, bells, tambourines and all, were accounted for by natural laws and that the so-called spiritualist was no spirit at all.

ELECTRICITY, ETC.

Robert-Houdin's employment of electricity, not only as a moving power for the performance of his illusions, but for domestic purposes, was long in advance of his time.

The electric bell, so common to us now, was in every-day use *for years* in his own house, before its value was recognized by the public.

When he fitted up his first call-bell, which he had done without the knowledge of his family, he fixed the stud beneath his table, that he might press it with his foot; and calling his children he said, " Here is a new trick.   When I put my finger in this tumbler of water, Adele will enter the room!"

And so it happened, for Adele was in the secret, and so on with his other servants.

As I have said, electricity was a force of highest value to him, and was an unknown factor among his professional brethren.   His inventions of mysteriously-moving clocks are numberless, and I lack time to describe them, but his application of electro-magnetism, long after he had quitted public life, demands a brief notice.

His dwelling-house in his retirement at St. Gervais, near Blois, stood about a quarter of a mile from the entrance gate, and when the traveler reached it, and used the knocker gently or forcibly, an immediate loud ringing became audible in the house, though so far away, which continued and would not cease its warning sound, until a servant pressed a stud placed in the hall, which immediately unlocked the gate, and an enameled plate appeared on it, bidding the visitor "walk in."

This gate, in opening and closing (the latter being done by the aid of a spring), set in motion at different angles of such opening and closing, a bell which rung in a particular manner, and the peculiar and quickly-ceasing sound of that bell indicated, with a little observation, whether the visitors were one or several in number, or a friend of the family, or callers for the first time, or a tramp.

His letter box, too, at the gate was a very ingenious contrivance. It closed by a small flap, which, directly the postman opened, set in motion an electric bell at the Priory. The postman had orders to put in, first, all newspapers and circulars, so as not to create unfounded expectations; after which he put in letters, one by one, so that in the house, if not inclined for early rising, he could, even in bed, reckon up the different items of the morning post-bag.

Then, to save the trouble of posting his letters in the village post-office (for Robert-Houdin wrote all his correspondence at night), by turning an apparatus called a commutator, the working of the signals was reversed, and the next morning the postman, on putting his parcel in the box, instead of causing a ring in the house, was warned by the sound of a bell close beside him, to go up to the house and fetch some letters, and he announced himself accordingly.

He had a favorite horse, named Fanny, for whom he entertained great affection, and christened her "the friend of the family."

She was of gentle disposition, and was growing old in his service; so he was anxious to allow her every indulgence, especially punctuality at meals, and full allowance of fodder.

Such being the case, it was a matter of great surprise that Fanny grew daily thinner and thinner, till it was discovered that her groom had a great fancy for the art formerly practiced by her master, and converted her hay into five-franc pieces. So Robert-Houdin dismissed the groom, secured a more honest lad, but to provide against further contingencies and neglect of duty, he had a clock placed in his study, which, with the aid of an electrical conducting wire, worked a food supply to the stable, a distance of fifty yards from the house. The distributing apparatus was a square, funn shaped box, which discharged the provender in prearrang quantities. No one could steal the oats from the horse aft they had fallen, as the electric trigger could not act unle the stable doors were locked. The lock was outside, and anyone entered before the horse had finished eating its oa a bell would immediately ring in the house.

This same clock in his study also transmitted the time two large clock faces, placed one on the front of the hous the other on the gardener's lodge, the former for the bene of the villagers.

In his bell-tower he had a clockwork arrangement of su ficient power to lift the hammer at the proper moment. Tl daily winding of the clock was performed automatically b communication with a swing door in his kitchen, and th winding up apparatus of the clock in the clock-tower was s arranged, that the servants in passing backwards and fo wards on their domestic duties, unconsciously wound up tl striking movement of the clock.

He had a marvellous contrivance for arousing his servan and compelling them to get up in the morning. The aları sounded, and continued ringing until they got out of bed t press a stud at the furthest end of rooms.

In addition to the foregoing, when for any reason h wanted to advance or retard the hour of a meal, by his mod of regulating the clock in the tower, he could, by secretl pressing a certain electric button in his study, put forwar or backward all the clocks as well as striking apparatus. I mystified the cook, and he had gained his purpose withou losing his character for punctuality at meals.

For a fuller account of these wonders see Robert-Houdin own description in "Le Prieure" and "Les Secrets de I Magie"

His grounds, very extensive and always maintained i strictest order, were so full of marvellous arrangements tha in the country round he had the reputation of possessin supernatural powers, which in the days of the Scotch witche recently spoken of before this brotherhood, might have co: him a good cremation! But during the Franco-Prussia war, and at the period when a descent upon Blois was b

íno means impossible, he was entrusted with every descrip-
ition of property by his confiding neighbors and friends; and
ihe constructed in an adjacent wood a cave, unknown to all
:the world, where he secreted these valuables until the crisis
was happily at an end.

ι It was at about this period that he sustained the severest
loss that had ever overtaken him, by the death of his young-
ιest son, Captain Eugène Robert-Houdin. His letter an-
ιnouncing that event may be of interest, so I reproduce it *ver-
ιbatim.*

"ST. GERVAIS, pres Blois, le 11 7bre 1870.

ι CHER MONSIEUR:—Je vous remercie bien, vous et votre
'famille, des marques de sympathie que vous m'avez ad-
'dressées au sujet du malheur qui m'a frappé.

ι Depuis la mort de mon pauvre enfant, je suis malade,
découragé et tout absorbé par ma douleur; c'est ce qui vous
explique le retard que j'ai mis à vous repondre.

Vous pourrez, mon cher ami, juger de l'etendue de mes
ιregrets par les details que je vais vous donner: Mon fils
avait 33 ans, il etait capitaine depuis 1866: il avait donc
quatre ans de ce grade; il faisait partie du 1er Zouaves et il
etait cite commes un des braves parmi ce brave corps. Vous
allez en juger par le recit suivant que j'extrais d'un article
du *Figaro* du 3–7bre, sous le titre de *Un episode de
Reichshoffen extrait d'une lettre particuliere.* Cette lettre
revient sans doute d'un soldat de la compagnie de mon fils;
elle est signée d'un X.

Je passe les détails navrants qui ont précédé cette malheu-
reuse retraite.   .   .   .   .

    .   .   .   "La ligne avait recu l'ordre de rompre et
nous étions vaincus, 35,000 contre 140,000!   On fit monter
de nouveau ma compagnie (1er Zouaves) sur le champ de
bataille, et l'on nous déploya en tirailleurs; seuls, sans artil-
lerie, nous devions soutenir la retraite.

"Ici commence un épisode de Waterloo.

"Sur l'ordre du capitaine Robert-Houdin le lieutenant
Girard s'avance avec deux hommes pour reconnaitre l'ennemi.
Il fait trois pas et tombe en disant 'n'abandonnez pas le
Coucou' expression familière par laquelle nous désignons le
drapeau.   Nous l'emportons et le Capitaine crie Feu!

23

"L'ordre de rétrograder nous arrive, mais nous ne l'enten-
dons pas et continuons à nous battre contre un mur de feu qu
éclaircit nos rangs. Bientot le capitaine tombe à son tou
en me disant: "Dites leur . . . . .
que je tombe le dernier en faisant face a l'ennemi." .

. . . . !
Une balle lui avait traversé la poitrine Transporté à l'ambulance de Reichshoffen, il y mourut quatre jours aprés, de
suites de sa blessure.

Eh bien! mon cher Manning, croiriez vous que ce brav
fils, au moment même ou il venait d'être frappé mortellemen
eut l'héroique courage, au milieu de la mitraille de tirer d
sa poche une carte et un crayon et d'écrire au dos ces mots
cher pere, *je suis blesse; mais rassure-toi; c'est un bobo.** S
signature n'a pu être achevée. La carte et l'envellope qui l
contenait sont maculées de son sang. Cette precieuse rel
ique m'a été envoyée de Reichshoffen apres la mort de mo
fils.

En voila bien long, cher monsieur, sur ce sujet. Mais j'a
pensé que ces details vous interessaient.

Veuillez me croire toujours,
Votre bien devoue,
ROBERT-HOUDIN."

TRANSLATION OF ABOVE LETTER.

"SAINT GERVAIS, *near Blois*, Sept. 11, 1870.

DEAR SIR—I thank you and your family for your token
of sympathy in my bereavement. Since the death of m
poor child I have been sick, discouraged and entirely ab
sorbed by my suffering; let that be the excuse for my dela
in answering.

You can judge, my dear friend, of the intensity of my re
grets by the following details.

My son was thirty-three years old; he was captain since
1866; he belonged to the 1st Zouaves and was considere
one of the bravest in that brave corps. You can judge of i
by the following extract from an article in the Figaro, o
Sept. 3rd, entitled "An episode of Reichshoffen," an extrac
from a private letter. This letter was undoubtedly writter
by a soldier in my son's company; it is signed with an X

*Expression qui designe en francais le moindre des maux que l'on puisse souffrir.

24

I omit the harrowing incidents which preceded this sad retreat. * * * *

"The line had received orders to break up and we were defeated, 35,000 against 140,000! My company (1st Zouaves) was drawn up on the battle-field, to be used as sharp-shooters, alone, without artillery; we were to resist the retreat."

Here begins an episode of Waterloo.

"Upon the order of Capt. Robert-Houdin, Lieut. Girard advanced with two men to reconnoitre the enemy. He took three steps and fell, crying: 'Do not give up the Coucou' (a familiar expression applied to the flag). We carried him away and the Captain shouted "FIRE!"

"The order to retreat came, but we did not hear it, and continued to beat against a wall of fire which illuminated our ranks. Soon our captain fell, saying: 'Tell them * * * * that I fell facing the enemy.' A bullet had pierced his breast. He was taken in the ambulance to Reichshoffen where he died, four days later, from his wound."

My dear Manning, would you believe it, my brave son, mortally wounded as he was, had the heroic courage amidst flying shot to take from his pocket a pencil and a card and to write these words:

'Dear father, *I am wounded, but be reassured, it is only a trifle!*' He could not sign this. The card and the envelope are stained with his blood. This precious relic was sent to me from Reichshoffen after my son's death.

I have written much on this subject, but I thought these details would interest you.

Believe me, your devoted,

ROBERT-HOUDIN."

Once more I feel that I must throw myself on your clemency, as I am about to show two articles of quite recent construction, and which, although exciting the admiration of all lovers of the marvellous, are in themselves mere toys compared with the minute and elaborate handiwork of the greatest of modern mechanicians. Two apologies would seem superfluous on the same subject; but I forgot to state that, when debating with myself whether I should produce any experiments at all in illustration of my discourse, the idea possessed me that there are three sorts of men (there are

many other sorts, of course) of whom I had to think in d
ciding this problem,

1. The men who know nothing. (Very few!)
2. The men who know something. (Happily, numerous
3. The men who think they know something. (No:
present to-night!)

Well, after a not angry discussion with myself, I dete
mined that if peradventure there were only ten innoce
ones among us to-night, (I know five among the Brethre
myself being one), *for the sake of those ten* I would show n
manifestations and demonstrations, and risk the cons
quences.

So here is a daintily modeled Guitar-Player. I exhil
this moving figure in order to make a comparison which w
not be odious exactly, but which may enable me to expla
the wonderful difference between the minute work of n
friend (whose sole aim was to imitate most closely the worl
of nature) and the limited movements of my Ethiopian Se
enader now playing before you. Houdin's Guitar-Play
not only moved its head, eyes and body in keeping with tl
air it was playing, but each of the tiny fingers touched tl
strings at the identical moment that the notes sounded fro
the concealed musical box at the base of the automaton.

My second modern figure I exhibit, as it is an excelle:
example of Houdin's Debureau, a French clown, who n
only came out of his own box and went through many pe
formances, but played an air on a small whistle placed in h
mouth, and finally smokes a pipe.

With his Oddship's kind permission, my French clow
will not only survey this distinguished assembly through h
glasses, but will, as you see, puff his cigarette after the mo:
approved fashion, and eject his long wreaths and rings t
perfumed smoke across the room till, so to speak, all i
blue.

Robert-Houdin's untiring industry manifested itself at a
early age, but the feat that established his indomitable wi
in overcoming difficulties, which to most enthusiasts woul
have appeared insuperable, was the successful imitation, bi
by bit, of a most delicate piece of mechanism, consisting o
a musical snuff-box (sent to his father for repair), fror

whose top a tiny bird sprang forth, singing its one sweet song, and then retreating to its hidden nest. This success, accomplished out of his regular business hours, gave him courage for further attempts of a still more ambitious nature, and during his brilliant career he was the inventor of numberless marvels of creative skill, all of them mysterious, all of them beautiful, and some of them absolutely poetic.

My last souvenir consist of this clock, one of his earliest inventions, which brought his name as a watchmaker into prominent notice, and which was commercially a great success. He was very anxious to be an early riser, but with the best resolutions he wanted (like other well meaning people we might mention) a good deal of awaking, and notwithstanding his loudest alarm, he was prone to turn round on the other side and go to sleep again, especially in dark weather. So this example of his own handiwork helped to cure him of his weakness, *by supplying him with a lighted match,* and a; the last tinklings of the alarm were dying away, the match was staring him in the face, he lighted his candle by it, got up, and went to his workshop or his study. You will perceive that at the proper moment the match, which had been previously placed in its receptacle horizontally, is rapidly drawn through two pieces of rough glass-paper, is lighted by the friction, jumps up to the vertical position, and insists upon being used for lighting the neighboring candle.

I intentionally omit, as being too long for this address, his adventures in Algeria, although they are intensely interesting, and I content myself by saying that in 1856 he accepted an engagement from the French Government to put an end to the belief among the Arabs in the miraculous power of their wizards and marabouts, whom he met on their own grounds, fought with their own weapons, and demonstrated under the public eye that he was more than a match for the best of them, though denying that he possessed any supernatural gift whatever.

And now I feel that my task is approaching completion—a pleasant task, but which I must not for your sakes make unduly long. I have not troubled you with many dates cr facts with regard to birth or history of my hero, one of the most remarkable artists of his time.

Had he lived till tomorrow, he would have been eigh
five years of age, and heaven only knows what new marv
of invention he would have given to the world!

I have endeavored, very rapidly, to give a sketch of
good friend, who was one of the most interesting of m
He had an individuality peculiarly his own. He had a geni
ity of manner positively magnetic, and exerting his inf
ence upon all who knew him.

His figure upon the stage was never to be forgotten. I
animation, his gesture, his ready wit, his quick transitio
from fun to serious earnest, would have fitted him for t
highest forms of acting—COMEDY and TRAGEDY wou
both have claimed him as their own!

He never played twice alike, and never flagged for a m
ment; but an interruption from a member of his audien
invariably drew forth some brilliant but good-natured rep
tee, which was crushing, for he was a fellow of infinite je
He was no common entertainer surrounded with showy sta
properties, for as Carlyle said of Dickens' readings, "h
face was the scenery!"

But, alas! the time came when the final trick was playe
and the final bow was made, and the inevitable curtain can
rolling down, and forever shut out the brilliant conjuror fro
a wondering and mystery-loving world.

The dead send no ambassadors to speak for them; but th
illustrious dead leave disciples behind to tell again the stor
and the glory of their lives. So tonight, as his humble di
ciple and reverent admirer, I offer to my Brother Odd Vo
umes this tribute to the memory and the genius of ROE
ERT-HOUDIN.

BIBLIOGRAPHY.

Robert-Houdin published the following works connected
/ith the arts and sciences, viz.:—

Les Confidences d'un Prestidigitateur. J. Hetzel. Paris.
858. 2 vols.

Les Tricheries des Grecs Devoilees. J. Hetzel. Paris.
863.

Le Prieure. Michel Levy freres. Paris. 1867.

Les Secrets de la Prestidigitation et de la Magie. Paris.
868.

Les Radiations Lumineuses. Blois. 1869.

Exploration de la Retine. Blois. 1869.

Magie et Physique amusante (oeuvre posthume). Paris.
877.

Most of the above works have been republished in English
y Messrs. Routledge and Sons.

30

To the Profession:

In presenting you with our latest catalogue, would call your attention to the trite saying: "Imitation is the sincerest flattery." Intending purchasers are therefore respectfully requested to bear in mind the fact that it is possible to copy an apparatus so that it has the appearance of genuineness, and offer it at a lower price than the original; and unless the customer is acquainted with all the details of construction and knows what the essentials are, he cannot detect that the work is made to agree with the price. Every care is bestowed on all apparatus leaving our hands, and each piece is first tried and proven to be perfect before it leaves us. Complete and explicit directions with each trick or illusion. Standard goods have a standard price. Look out for people that have something just as good for a good deal less.

While we make any trick or apparatus not mentioned in this list, we hold in reserve at all times a number of sensational effects suitable for artists in all branches of the profession.

Perfect instruction in the higher branches of modern Parlor and Stage Magic on reasonable terms. When outfits are furnished instructions are gratis.

This entire catalogue is copyrighted and any infringements on the same will be prosecuted to the full extent by our attorneys, the Hon. A. S. Bradley, and Messrs. Arthur and Boland, of the Ashland Block, Chicago.

Soliciting your favors which shall have prompt attention, we are

Fraternally yours,

CHAS. L. BURLINGAME & CO.,

Studio and Address for Telegrams,   **Box 851 CHICAGO, ILL.**
5766 La Salle St., Chicago.

TERMS:

On small orders, one-quarter cash with order, balance C. O. D. On large orders, one-half cash. Orders under five dollars not sent C. O. D., must be fully prepaid. Telegraph orders ignored unless a deposit is made.

BOOKS.

"Of all those arts in which the wise excel,
Nature's chief masterpiece is writing well."

MODERN MAGIC.

A practical treatise on the art of Conjuring. By Professor Hoffmann, 528 pages... $1 5

DRAWING ROOM AMUSEMENTS

and evening party entertainments. By Professor Hoffmann, 512 pages... 1 5

THE SECRETS OF CONJURING AND MAGIC,

Or how to become a wizard. By Robert Houdin. Translated and edited with notes by Professor Hoffmann, 394 pages..... 2 5

THE SECRETS OF STAGE CONJURING.

By Robert Houdin. Translated and edited with notes by Professor Hoffmann; 252 pages............................... 1 1

DRAWING ROOM CONJURING.

Translated and edited with notes by Professor Hoffmann, 192 pages.. 1

MORE MAGIC

By Professor Hoffmann, 457 pages, 140 engravings.............. 1

SLEIGHT OF HAND.

A practical manual of Legerdemain for amateurs and others. Illustrated, by Edwin Sachs............................... 2

LETTERS ON NATURAL MAGIC,

by Sir David Brewster. Illustrated. Contains full explanation of the automatic chess player........................... 1

THE ART OF MODERN CONJURING, MAGIC AND ILLUSIONS,

Thought Reading, Mesmerism, etc., etc. Illustrated, by Henri Garenne (Professor Lind.)................................ 1

"SHARPS AND FLATS,"

a complete revelation of the secrets of cheating at games of chance and skill. Illustrated, by John Nevil Maskelyne..... 1

"LEAVES FROM CONJURERS' SCRAP BOOKS,"

Or Modern Magicians and Their Work, by H. J. Burlingame. Contains all about Hypnotism, Mind Reading, Second Sight,

Instantaneous Memorization, Foreign and American Conjurers, Prominent Amateurs, The Herrmanns and Harry Kellar. Anna Eva Fay's Chicago Experience. Interesting Reminiscences. Explanations of the Cocoon, Growth of Flowers, Several Cremations, Amphitrite, Spirit Bell, "Mystery of She," and many other tricks and illusions. Cloth and Gold, 274 pages, fully illustrated. A standard work. Price............ $2 00

"AROUND THE WORLD WITH A MAGICIAN AND A JUGGLER."

Unique Experiences in Many Lands. By H. J. Burlingame. From the papers of the late Baron Hartwig Seeman, "The Emperor of Magicians," and William D'Alvini, Juggler, "Jap of Japs." Contains valuable and rare information for the profession, contracts and programmes, giving the renowned D'Alvini's entire programme, also life sketch of the celebrated Bellachini and the trick that made him famous, together with the celebrated essay of Dr. Max Dessoir on "The Psychology of the Art of Conjuring." Cloth and Gold, 172 pages, fully illustrated. Price................................... 1 00

"HERRMANN; HIS LIFE; HIS SECRETS."

By H. J. Burlingame. This handsome volume is just such a book as will delight and instruct the professional and amateur. One-half of the book is devoted to the history of Herrmann the Magician, his family and the career of his famous brother, Carl Herrmann. Then follow full descriptions of over fifty of the tricks that have made the name of Herrmann famous, and the equal of that king of conjurors, Robert-Houdin, with forty-three illustrations, portraits and halftones. The author of the work has devoted twenty-five years to inventing, manufacturing and selling many of the most popular magical apparatus made in this country, and is consequently able to write upon these subjects with peculiar interest. Among many of the most curious revelations in the book is a complete elucidation given for the first time of the New Marvelous Lightning Thought Transference, recently performed here by Kennedy and Lorenz. Beautifully bound in Holliston cloth, rough edges, polished red top, fancy designs. 300 pages. Price,................................... 1 00

See Table of Contents on next two pages.

TABLE OF CONTENTS

The Last Program of Herrmann the Great in Chicago, January 15, 1896.

"TRICKS IN MAGIC, ILLUSIONS AND MENTAL PHENOI ENA."

Volume I. Compiled by H. J. Burlingame. Containing explanations of the following: Thought Transference and Clairvoyance, as used by Kellar, Morritt, Berol & Belmonte and others; Tachypsychography or Long Distance Second Sight, : Psychognotism as used by Guibal and others, Hypnognotism, Second Sight Through Brick Walls, Spirit Thinkephone, New Silent Second Sight, Head of Ibykus, Sing Sing Mystery, Mango Tree, Great Shooting Act, Noah's Ark, Oriental Barrel Mystery, Great Mahatma Miracles, Spirit Circles under Test Conditions, Bank Note Tests, Rope Tying Feats, Many Illusions, Edgar Poe's Raven in the Garland of Thebes, Samuel's Mystic Percolator, Samuel's Magic Squeezers, Wonder Kraut, Wine Tassels, etc., etc. Sixty-three effects in all. Price.. $0 :

"TRICKS IN MAGIC, ILLUSIONS AND MENTAL PHENON ENA."

Volume II. Compiled by H. J. Burlingame. Containing explanations of the following: Eglinton's Famous Slate Trick; Sealed Letter Reading; the Spirit Rapping Decanter; the Reading of Folded Papers; Blood Writing on the Arm; Reading Cards Blindfolded; Thought Reading in Cards; Samuel's Vanished Mirror and Spectral Demon; the Latest Slate Mystery; the Winged Numbers; Yank Hoe's Cigarette and Card Trick; Yank Hoe's Paper Trick; Verbeck's New Dictionary Trick; Ornithological Labyrinth of Perplexity; Tambourine and Paper Trick; Candle and Rings; Transmigration of Smoke; Indian Illusion with Rings; New Vanishing Pocket Knife; Instantaneous Bouquet Production; the Blackboard Feat; the Celebrated Bank Note Test; the Maid of Athens; Vivisection and all those tricks known as Valensin's Tricks or Inventions. Price.. O :

"TRICKS IN MAGIC, ILLUSIONS AND MENTAL PHENON ENA."

Volume III. By H. J. Burlingame. A very important volume, just off the press. Contains explanations of the following three tricks by the well-known writer, Prof. Hoffmann. "The Magic Tambourine," "The Great Dictionary Trick," and "The Climbing Ring." Also Maskelyne's "Spiritualistic Couch," "The Revolving Bust Illusion," "Buatier's Human Cage," "Buatier in a Fix," "Morritt's Cabinet," "Denstone's Metempsychosis," "David Devant's Flying Thimble," "McLaughlin's Patented Thought Reading Trick," and "Euclid Outdone, or "The Spirit Mathematician." All fully illustrated with from one to five illustrations to each trick or illusion. In all twenty-eight illustrations.

These illusions are thoroughly described for the first time, and nearly all of them entirely new to American Conjurers.

A sketch of the career of Frederick Bancroft, the American Conjurer, with reason for his non-success is given.

The book contains also a complete Bibliography of Magic, Conjuring and Amusements, in English, German and French, being the most thorough work on this subject ever published. The Bibliography alone makes this a work of inestimable value to the professor, amateur and general reader. 378 works listed. Price.. $0 25

HYPNOTISM AS IT IS.

A Book for Everybody. By X. La Motte Sage, A. M., Ph. D., LL. D. Formerly Professor in Pierce College, Philadelphia, Pa., and Professor in Central College, Sedalia, Mo. Richly illustrated by 20 full page Photo-engravings. Dr. Sage has personally hypnotized over 10,000 people. He tells the public what experience has told him. Universally pronounced the best and most attractive book of the kind that has ever been published. Paper 8vo., 116 pages. Price............................. 0 30

THE REVELATIONS OF LULU HURST, THE GEORGIA WONDER.

Written by herself. Explains and demonstrates the Great Secret of her Marvelous Powers. A new and unparalleled revelation of the Forces that puzzled and mystified the entire continent. *Every test illustrated with full page half-tone engravings, and every one who reads the book can acquire the power.* Paper, 267 pages. Price,.. 0 50

ISIS VERY MUCH UNVEILED.

A story of the Great Mahatma Hoax. By Edmund Garrett. Contents: Part 1. The story of the Great Mahatma Hoax. Introduction. No Mahatma, No members. Mystification under Madame Blavatsky. The Psychical Research Exposure. Mystification under Mrs. Besant. Enter the Mahatma. Every Man his own Mahatma. The Adventures of a Seal. The Climax of Theosophic Brotherhood. The Mahatma Tries Threats. Mrs. Besant's coup de main. A meeting of the (Theosophical) Pickwick Club. Questions and Challenges. Part 2. From Officials. From Prominent Theosophists. From Private Members. Part 3. A General Rejoinder. Last shreds of the Veil of Isis. Postscript. Mr. Judge's Mahatma at Bay. L'Envoi, "The Society upon the Himalay." A reply from Mr. W. Q. Judge. An Appreciation of Mr. Judge's Reply. Illustrations and Fac-Similes. Frontispiece

Portrait of Madame Blavatsky. Portrait of Mrs. Besant. Portrait of Colonel Olcott. The "Mahatma Seal." The Envelope Trick. Fac-similes of Mahatma's Missives, of Mr. Judge's Handwriting, etc. Portrait Cartoon. "When Augur meets Augur." Paper.................................. $0 3

"THE DEATH BLOW TO SPIRITUALISM."

Being the true story of the Fox sisters as revealed by authority of Margaret Fox Kane and Catherine Fox Jencken. By Reuben Briggs Davenport. This book is, in fact, what its title sets forth. "The Death-Blow to Spiritualism." Paper, 274 pages, with vignette half-tone cuts of Margaret Fox Kane and Katie Fox Jencken. Rare and interesting work. Price,.......... 0 5

"CONJURING FOR AMATEURS AND PROFESSIONALS."

A Practical Treatise on How to Perform Modern Tricks, by Ellis Stanyon, F. O. S. The latest book on Magic by a popular and well-known professional. Contains many new, original and highly interesting tricks, including full directions and sketches for the famous "Paper Folding Act." Paper, 122 pages, illustrated. Price,.................................. 0 5

CONJURING WITH CARDS.

By Prof. Ellis Stanyon, F. O. S. A practical treatise on how to perform Modern Card Tricks. The latest English work. London, 1898. Paper 8vo. Illustrated, an excellent manual. Price,.. 0 5

"REVELATIONS OF A SPIRIT MEDIUM:

Or, Spiritualistic Mysteries Exposed." Written by a prominent medium of twenty years' experience. This is a rare and valuable work, and should be in the hands of every conjuror, medium or investigator. It is intensely interesting to the general reader. Paper, 324 pages, illustrated. Former price, $1 50; now,.. 0 7

LETTERS ON DEMONOLOGY AND WITCHCRAFT.

By Sir Walter Scott, Bart. With an introduction by Henry Morley, LL. D., Professor of English Literature at University College, London. Third Edition, Cloth, 8vo, 320 pp., scarce. Price.. 0 7

THE PSYCHOLOGY OF ATTENTION.

By Th. Ribot, Professor of comparative and experimental Psychology at the College De France, Editor of the "Revue Philosophique." Interesting only to students of Psychology. Price.. 0 7

"THE MODERN WIZARD."

By A. Roterberg. Contains eighty-four modern tricks. Cloth, 117 pages. Illustrated. Price.............................. 1 0

PRACTICAL PALMISTRY,

Or Hand Reading Made Easy. By Comte C. De Saint-Germain. Issued a few months ago, this volume has met with immediate recognition and very large sales. The press has declared it a standard. Illustrated with 55 pictures of hands. Special edition (with several additional half tones), bound in extra cloth, rough edges, polished red top, stamped with special design........................ $1 00

"LATTER DAY TRICKS."

Uniform with "The Modern Wizard." By A. Roterberg. 104 pages, containing seventy-five modern tricks. Cloth. Price   1 00

"NEW ERA CARD TRICKS."

By A. Roterberg. Devoted solely to Card Tricks, with and without apparatus. The most complete book on Card Tricks in the English language. Cloth, 284 pages, 203 illustrations. Price...   2 00

"HOURS WITH THE GHOSTS,

Or, XIXth Century Witchcraft. By H. R. Evans. The pretensions of the so-called Clairvoyants, Mind Readers, Slate Writers, etc., graphically exposed. The true story of Madame Blavatsky given to the world, with new and exhaustive evidence. A most conscientious, extraordinary work, convincing to a degree, and readable throughout. Exposes Slade, Keeler and others. Fancy cloth, 302 pages. Finely illustrated. Price.....................................   1 00

"HYPNOTISM, MESMERISM, AND THE NEW WITCH-CRAFT."

By Dr. Ernest Hart, of W. London Hospital. A new enlarged edition, with chapters on "The Eternal Gullible," the Confessions of a Professional Hypnotist, and notes on the Hypnotism of Trilby. Cloth, 212 pages, illustrated. Very interesting. Price..   1 50

"THE BOTTOM FACTS CONCERNING THE SCIENCE OF SPIRITUALISM."

By John W. Truesdell. Derived from careful investigations covering a period of twenty-five years, with many descriptive illustrations. This is the famous work on "Bottom Facts," out of print for the last ten years. It is one of the most valuable books for mediums or conjurers ever issued. Among other effects, it contains the complete spiritual work or programme of the Eddy Brothers, Dr. Henry Slade and Anna Eva Fay. Cloth, 331 pages, fully illustrated. Price.........   2 00

A THOUGHT-READER'S THOUGHTS.

Being the Impressions and Confessions of Stuart Cumberland.
Contains the travels and experiences of the celebrated Mind
Reader. Large 8vo, 326 pp. Fancy cloth and gilt, with
photographic portrait. Out of print and scarce. Original
price, $2 75. Our price.................................... $2 50 .

FIFTY YEARS IN THE MAGIC CIRCLE.

Prof. Blitz' original work. Being an account of the author's professional life, his wonderful tricks and feats, with laughable
incidents and adventures as a magician, necromancer and
ventriloquist. By Signor Blitz. Illustrated with numerous
engravings and portrait of the author on steel. Large 8vo,
432 pages. Very scarce and interesting. Price,.............. 3 00

"THE PRACTICE OF PALMISTRY FOR PROFESSIONAL PURPOSES."

By Comte C. De Saint-Germain, A. B., LL. M. (of the University
of France), President of the American Chirological Society
(incorporated), and of the National School of Palmistry. With
an introduction by the late Adrien Desbarrolles. Over 1,100
original illustrations and complete Palmistic Dictionary. As
a bread-winner orthodox palmistry is proving a great success,
and "The Practice of Palmistry for Professional Purposes" is
the only work of its kind that places in the student's hands a
money-making instrument of incontestable value. For the
first time, real, useful palmistry can be learned without a
teacher, for every word contained in this great book has been
tested upon class after class of intelligent pupils and produced the most complete and satisfactory results. *With this
a successful business can be commenced at once.* For the standing of the Art we need only cite you the career and success of
Heron Allen in this country and abroad, and Cheiro the
Palmist (Count Leigh de Hamong), now holding receptions
and readings at the Auditorium Hotel, Chicago. Owing to
our cordial business relations with the author and proprietor
of this monument of patience and learning, we have been ap-
pointed special sales agent for a limited number of copies only,
each one numbered by hand and with the author's autograph
signature. There are but few copies issued, and only one
copy will be sold in one town or city. The binding is of dark
red silk cloth, with back and corners in Russia leather and
edges full gilt, a superb and most practical makeup for a work
of exceptional value to the student. We cannot commend
the above too highly. Volume I, with portrait of Emma Calve,
price, $3 50. Volume II. $4 00. Both volumes in neat box,   7 50

The following article over the signature of the well-known scien-
tific writer Frederick Boyd Stevenson, appeared in The Journalist

("Devoted to All Who Make or Read Newspapers") of New York City, Vol. XXIII, No. 16, August 6, 1898.

"Magic, in its original sense, meant light and knowledge. Then the term was looked upon with suspicion, and magicians, although still deemed wise men, were regarded as sorcerers. The word 'magic,' in the modern acceptation, is a synonym for 'legerdemain,' and thus we confound the terms magician, conjurer and prestidigitator. There is always a certain fascination about that word 'magic,' whether we use it in the sense that it was originally used or apply it in the common understanding of the term—that is, an amusing trick performed for spectators. The modern magician is not the ancient alchemist searching patiently for the philosopher's stone, nor the meditative astrologer, nor yet the supernaturally wise physician. He is up-to-date. He knows all the latest tricks with cards and can give the Hindu conjurer, with his growing shrubs and headless boys, inside information. The Press Club of Chicago happens to have among its members just such a magician as this. Mr. H. J. Burlingame can do any of the astonishing things that all these other wonderful magicians have done. He, however, has looked into the practical side of magic, and not desiring to hide his light under a bushel has given his knowledge to the world.

Mr. Burlingame was formerly a newspaper man. He became interested in legerdemain, made the acquaintance of the most celebrated men in that profession, and was himself an expert. He has written several books on the subject that have been very popular and have had a large sale. In his book, 'Around the World with a Magician and a Juggler,' he tells the story of the life of Baron Hartwig Seeman, and gives an exceedingly interesting sketch of D'Alvini, whose real name was the unromantic Peppercorn. His book, 'Herrmann the Magician,' is full of good stories concerning the two who, in recent times, have borne that name — Carl and Alexander — and ;whose ancestor was almost equally as famous. He also has something to say of Cazeneuve, Kellar, Vanek, Heller, Samuels and Robert-Houdin, with whose works he was perfectly familiar and many ;of whose tricks he has plainly explained. Mr. Burlingame hates shams, and he scores unmercifully the clairvoyant, the mind reader and spiritualistic miracle workers who pretend to invoke the aid of the supernatural in their exhibitions. One of his most entertaining books, 'Leaves from a Conjurer's Scrap Book,' deals with these subjects in a way that shows what the author thinks of such people without any room for misunderstanding. As mind reading had its origin in Chicago, the chapter on that subject is especially interesting to residents of this city. J. Randall Brown was the first person who made a display of his so-called power. He had made a wager with an old resident of Chicago that he could find a pin, no matter where it was concealed, if it were placed in walking distance. The pin was placed beneath a rug in front of the Sherman House. Brown was blindfolded and led his friend there, winning his wager and becoming famous. He made a tour of the states and afterwards went to Europe. Washington Irving Bishop, whose death, while presumably laboring under the mental strain that attends mind readers, occasioned such a stir a few years ago, was an assistant of Brown.

Mr. Burlingame was born in Manitowoc, Wis., June 14, 1852. His father and mother were among the earliest settlers of Chicago. From here they went to Manitowoc in a wagon. After living in Madison for some time they removed back to Chicago. When he was about twenty Burlingame went to Rotterdam, Holland, and entered the commercial business, subsequently traveling through Germany and Switzerland, on foot, as correspondent for American papers. He remained abroad for a number of years, living for a time with his uncle, Pere Hyacinthe. After having lived in Chicago for some years he returned to Europe and made a professional tour as a conjurer. Then he came back to America, residing in Baltimore and Cincinnati, finally locating in his old home, Chicago. Mr. Burlingame now occupies himself almost entirely with magic. That he has not forgotten his old cunning as a prestidigitator the members of the Press Club can attest."

WITH CARDS.

1—MESMERIZED CARDS.

They cling to the palm of your hand and will not fall off, excellent impromptu trick......................................$ 10

2—CHANGING FACE CARDS.

Price.. 25

3—FINE FORCING CARDS.

Per pack of 28.. 25

4—FINE FORCING CARDS.

Per pack of 52.. 50

5—TORN CORNER CARD.

Price.. 40

6—MOVING PIP CARD.

Price.. 40

7—CARD CHANGING TO A ROSE.

Price.. 40

8—CARD THAT RISES IN PACK AND TURNS AROUND.

Price.. 50

9—THE CRŒSIAN CARDS.

A new feature in cards, as they produce, multiply or vanish coins. Very useful. Per pair........................... 50

10—IMPROVED NAILED CARD SHOT.

The trick of throwing a pack of cards against the ceiling or a door, and the card previously selected remaining nailed there, is well known. This improvement consists in borrowing a ring which is found fastened by a ribbon to the nail, through the card, although the lady from whom the ring is borrowed believes she holds the ring all the time....................... 25

11—READING CARDS BLINDFOLDED.

Performer is blindfolded, and on any pack of cards being placed in his hands, he at once proceeds to name each one in regular order, and allows cards to be shuffled at any time. A most excellent trick. Not to be had elsewhere............. 25

12—LA HOULETTE, *a la Buatier*.

Buatier uses no frame or case, but merely places cards in a tumbler, and by blowing through a paper cone the chosen cards jump out of the tumbler, and finally miniature cards of the same suite as selected ones appear inside cases of watches previously borrowed, but not opened; splendid effect; no apparatus required for this deception; secret.................... $ 25

13—NEW CARD AND CIGARETTE TRICK.

A selected card is torn in pieces, and pieces placed in an envelope, minus a corner of the card which is held by person who drew it, and who can also hold the envelope. Artist then borrows a cigarette, and finding it will not light, scrapes the paper off, and discovers in it the card, restored, except missing corner piece, which fits exactly, and on opening envelope the tobacco is found in place of the pieces of torn card......   25

14—CLAIRVOYANT CARD TRICK.

Three persons think of a card each, and performer by clairvoyance, writes down their names on three slips of paper and puts them in a glass; they are then asked to remove their cards from the pack, which is counted and of course the three cards are missing. Pack is placed on another glass, and the three selected cards in a card box from which they vanish, and on recounting the pack it is found that the three cards have gone into it, and on opening the papers the names written thereon are found correct. (NOTE—These cards are not forced, for the persons are only asked to think of them.) A splendid trick; the only apparatus required is a card box, which performer can easily make. New. Secret....................   25

15—DEMON CARD TRICK.

A very effective trick. Three cards are selected by audience, previous to which three pieces of blank paper are given for inspection, and placed on a slate in full view. Cards returned to pack, and papers opened, when names of cards are found written on the papers. This is a first-rate deception, and new. No apparatus required. Very easy..................   25

16—OUR SPELLING BEE.

To spell or lay out names, numbers and suites of cards, and days of the week, with names of months. Entirely new and first time offered for sale. As introduced by us with great success in England, Holland and Germany. Any pack of cards is used (the puzzling eight-card "lay out" included)............   25

17—THE DEMON CHANGING CARD.

Genuinely changes three times in the very midst of audience. There are cards made to change four and six times, but this

is absolutely the very best, for the reason that the card is
given for inspection.   New............................... $ 50

18—TO CHANGE A HANDKERCHIEF INTO A PACK OF CARDS.

Entirely new and striking, free-hand work, very useful and easy.
. Complete...... .... 1 00

19—BLINDFOLD PRODUCTION.

A pack of cards are shuffled and placed in performer's coat
pocket, he is blind-folded and yet produces from pocket at
once, any card called for by the audience. Secret..........   25
Complete with cards....................................... 1 00

20—NEW THOUGHT READING IN CARDS.

A sealed envelope is first handed to a lady, and one to a gentleman. From among four packs of cards the gentleman himself selects one pack and the lady writes on a slip of paper
the name of her favorite flower or bird. The envelopes are
then opened and found to contain a written or printed list of
just what the lady and gentleman selected. Novel and very
interesting, no confederacy. Secret only...................   25

21—STRIPPERS WEDGERS AND BISEAUTE CARDS.

For doing all tricks, fine quality, round corners. Per pack........ 1 00
Same, second quality.......................................   50

22—ELECTRIC AND CASCADE CARDS.

For fancy and expert shuffling, unequaled. Per pack............ 1 00

23—DISSOLVING AND VANISHING PACK.

Performer, after illustrating several tricks with an ordinary pack,
takes several cards, and without covering them in any way, or
turning his back, causes the cards to dissolve before the eyes of
audience until they are about half their original size; they still
"grow smaller and beautifully less," until now they are almost
as large as a postage stamp. Performer holds the little pack
now on the tips of fingers, when, to heighten the effect to a
greater summit, they vanish, and performer shows both hands
empty. A very fine trick. Complete....................... 1 00

24—NEW UNEQUALED CARD PASS.

To at once produce from a spectator's pocket a number of cards,
although hand is shown empty just before inserting in pocket.
A new and striking effect. For clever performers only......   25

25—THE MYSTIC REAPPEARING CARDS.

Performer hands to any number of audience a small piece of ordinary paper and requests him to make a small packet or envelope, about the size of a playing card, and seal it with wax which is furnished. Any person now selects three cards from a pack; cards are returned to pack, shuffled, and placed in spectator's pocket. The folded paper is now opened by performer and found to contain the three selected cards, and on person examining pack, he has, he finds, the three cards have disappeared from it. No duplicate cards used. For clever performers. No assistant required. Fine effect............. $ 50

26—THE ENCHANTED SHOT.

New card target, own invention. Very useful and unequaled. Any selected card is impaled on same by a bullet or a dart, from a blow-gun or a pistol. This is quite a departure from the old style targets. Done anywhere and a fine novelty..... 1 50

27—NEW "CARDS ON CHAIR."

Any chair used in performing this trick: Request three persons of audience to select one card each; they replace same and pack is shuffled. At a word from artist the three chosen cards instantly appear on the back of chair................. 2 50

28—NEW JAPANESE CARD TRICK.

Artist shows an ordinary Japanese fan and then allows three cards to be selected which are either loaded in pistol and shot at the fan, or are simply thrown at it, when all three cards are caught on the fan, and can be at once removed for closer identification. New, novel and striking...................... 2 50

29—NEW HOULETTE AND CARD ILLUSION.

Show an ordinary pack of cards, from which request several ladies to draw a card each, usually three or four; pack has been given for examination. Now bring forward houlette and also give that for examination. When houlette is returned to you place pack in it and request first lady to place her card in anywhere she likes, not letting you see the card nor touch it. Now place handle of houlette to your mouth, command card to rise, and the audience see the card gradually rising. Same request is made of the others who drew cards, and in same manner they are caused to rise. Cards can be made to rise slowly or quickly or jump out of houlette. You can also hold houlette in hand and at arm's length and command cards to rise. New and very effective 3 00
Same in beveled plate glass, nickel-plated................ 5 00

30—CARD IN LIGHT.

The effect of this trick must be seen for the beautiful mechanism to be appreciated. One of audience selects a card from pack; the card is torn up and pieces placed in a pistol. A lighted candle in a candlestick is now placed on table or held in hand of artist. One of company takes pistol and fires at candle, when instantly the chosen card appears in place of flame of candle, and can at once be taken off and shown as the real card. All sides of candlestick shown and may be used anywhere $3 00

31—THE ECLIPSED CARD.

The performer allows any person to select any card from the pack and allows same person to replace the card in the pack and shuffle same. This person can then look through the entire pack of cards and will find that the selected card has totally disappeared. Card can reappear anywhere desired. No particular skill required. Entirely new. Complete with cards 2 00

32—NEW DISSECTING CHANGING CARD BOX.

This is a skeleton dissecting changing card box, that does all the work of the old style flap box, but infinitely superior; can be examined and not detected; puzzles professionals. Described and illustrated in Ellis Stanyon's work on "Conjuring with Cards." Each.. 2 00
Per pair... 3 50

33—RESTORED CARD ON HAT AND BOTTLE.

A selected card is torn or cut to pieces by person drawing it, who retains one piece of card, the remaining pieces performer loads into a pistol and allows a gentleman to fire same at a hat, when the card appears on the hat instantaneously, minus the missing portion, which fits exactly. The mutilated card is now placed upright in a cork in a bottle and the missing portion of card placed in pistol, which is now fired at the card and same immediately appears fully restored and is taken off and handed for examination, as is the bottle and cork. Very effective. Price, including cards.......................... 1 50

34—THE FLOATING HOULETTE.

A delicate nickel-plated houlette is suspended above the centre of stage by ribbons secured to each side of houlette, the other ends of ribbons extending to the sides of the stage where they are also secured. After a number of cards have been selected and returned to pack, it is placed in the houlette, when the drawn cards rise from same, one at a time, making a most striking effect. This was a popular feat of the late Alexander Herrmann... 5 00

35—THE CARD SWORD.

This very interesting feat commends itself to every magician—a trick that has always received the enthusiastic applause of those who have had the pleasure of seeing it well executed.

Three cards are selected by audience and replaced in pack. Performer, who appears with an elegant sword, invites some person to throw the pack into the air. Instantly he does so performer makes a lunge with his sword among the flying cards, and succeeds in thrusting his keen edge through the three chosen cards, which are given to audience for identification. *This is an entirely new make of sword*—all sides shown; can be carried without fear of detection into midst of audience —and is of elaborate finish *with our original improvement.* Price.. 4 00

36—MECHANICAL RISING CARDS A LA HARTZ.

Cards are placed in a tumbler and it is held by one of the audience in midst of audience, and cards rise as called for. Noiseless and elegant... 10 00

37—NEW RISING CARDS, "NONESUCH."

A skeleton card frame, only large enough to hold a pack of cards, consisting of two sides, no back or front, is handed for examination, then placed on top of a delicate stand, scarcely one inch thick, no fringe or drapery. After any number of cards are selected, the pack is placed in the skeleton case, and the chosen cards rise as desired. The pack can be taken out and shuffled at any stage of the trick. No previous preparation of cards necessary. Complete with cards, including torn card restored, the moving pip card, and the card that comes up back to the front and turns around. New and very fine... 8 00

38—STANSELLE'S PNEUMATIC RISING CARDS.

The latest and best rising cards. An ordinary glass bottle is placed on stand or table; on the bottle a plate glass houlette. After selected cards are replaced in pack by audience, the pack of cards is placed in the houlette, when the cards rise on command, the last one jumping out. No threads, strings or assistant required. Entirely new and a great surprise to any audience... 10 00

39—CARDS UP THE CHAIR.

A SENSATIONAL RISING CARD FEAT.

A small chair, twenty inches high, is examined and placed on table. Three selected cards are torn up and loaded in pistol; the remainder of cards are placed inside the seat of chair which serves as a holder. At report of pistol the pack of cards rise in a column against the back of the chair. Now at command of the artist, the chosen cards rise one at a time from the top of the column, one rising upright and the other two projecting sideways. The chosen cards are removed, and at command column descends, when all cards and chair are handed for examination. New and original, first-class effect......... 10 00

WITH COINS.

40—FLYING COINS THROUGH CARDS.

Any pack of cards is placed on any ordinary tumbler or goblet, and three borrowed half dollars are at once passed through the cards and are seen to fall into glass, while the performer stands at a distance from same. Very effective..............$ .25

41—HERR DOBLER'S COIN AND ORANGE FEAT.

He borrows a coin and has it marked by audience; shows same to second gentleman, and tells him also to mark it. Two oranges are on a tray on side table. Then he holds coin in right hand and asks audience which of the two oranges they desire him to pass the coin into. Suppose they say "left," he instantly vanishes coin, and requests one of audience to cut open orange with ordinary knife, when *the marked coin* is found therein. Now he says, "You fancy that if you had chosen the other orange I should have failed, but it is not so, as I shall illustrate." With such words he, taking up the coin, commands it to travel into centre of second orange, which, when cut open, is found to contain the *marked coin.* Coin given to audience to test its genuineness after each pass. No apparatus required. No mechanical knife used. Secret............... .25

42—NEW AND MARVELLOUS COIN PRODUCTION.

This trick consists of producing any quantity of coins, gold, silver or copper, from the pockets of various members of the audience. A volunteer having been obtained, the performer shows his right hand unmistakably empty, and forthwith plunges it into the pocket, immediately bringing forth a handful of the required coins, which he lets fall in a stream into a borrowed hat held in the left hand. The operation is then repeated with a second spectator, and so on as often as desired, the supply being apparently inexhaustible. Very fine for clever performer, secret all that's needed.................. .50

43 A—NEW STYLE COIN HOLDERS.

These are very fine indeed for catching money in the air. Used by skillful performers. Each.......................... 1 50

43 B—COIN DROPPERS FOR STAGE ONLY.

These are easily arranged in the flies or attached to the chandelier, and drop coins visibly into a borrowed hat. Each.......... 2 50

43 C--EXCELSIOR COIN HOLDER FOR MONEY CATCHING.

The invention of A. G. Waring. Hands shown perfectly empty
when borrowing a hat, yet coins are caught just the same,
"slick as a feather." Hands shown empty at any time. Never
offered before. First-class................................... 1 50

44--THE DEMON HALF DOLLAR.

A half dollar having been examined by audience is returned to
performer, who, before the very eyes of company, causes it to
multiply into two. The two coins are now laid on back of
hand, and rubbed by finger back into one. Easily manipu-
lated, and a good mechanical trick........................ 2 75

45--THE DEMON DOLLAR.

Same as No. 44, but dollars are the coins used................ 3 50

46--WANDS FOR CATCHING COINS.

A. Best wand for catching real half dollars, including dummy
wand and money slide (dropper for vest)................ 5 00
B. New style fine half dollar catching wand without tips........ 4 00
C. Fine wand, with imitation half dollars, good enough you will
say.. 3 50
D. Wand for catching quarter dollars......................... 2 00
E. Wand for catching two-cent pieces....................... 75

47--A NEW MINT.

A very fine, small piece of apparatus, easily palmed, and which
changes three borrowed half dollars to copper, copper to
brass, and the brass to half dollars again. Entirely new, and
very useful for free hand work in changing vanishing or producing coins; (half dollar sizes only)...................... 1 50

48--THE MESMERIZED COINS.

To balance three coins on their milled edges, one on top of the
other. Nice pocket trick. With coins.................... 75

49--MULTIPLYING COIN PLATE.

Prof. Herrmann's Detroit Banker. To produce or multiply coins.
Nickel-plated or China.................................. 1 50
Same, extra large...................................... 3 50

50--THE BLUE GLASS FOR COINS.

An old timer, but good; produces four half dollars, without covering, after being shown empty............................ 3 00

51--NEW COIN CHANGING TRAY.

This is a small handsome nickel-plated tray which changes coins
without covering; very useful in combination and in securing
marked coins.. 3 50

52—NEW NE PLUS ULTRA MONEY CATCHING FEAT.

Performer borrows a hat. He catches twenty-five or thirty coins—half dollars or dollars—out of the air and drops them visibly from finger tips, one at a time, into the hat; after which he walks among the audience and asks some lady to hold out her hands. He then turns the hat over to pour the coins into the extended hands, but all the coins have vanished. Fine effect. Always produces a great deal of amusement. Something entirely new... 5 00

53—GLASS COIN FRAME.

A striking feat! Four borrowed half dollars are thrown, one at a time, towards a magnificent gold frame surrounding a sheet of glass, and are seen to alight on glass, one at a time; then, suddenly, all four fall at once into borrowed hat, or any other receptacle. Can be done anywhere. Very fine............... 10 00

WITH HANDKERCHIEFS.

"THERE'S MAGIC IN THE WEB OF IT."—*Othello, III, 4.*

54—THE DISSOLVING HANDKERCHIEF.

To vanish instantly any borrowed handkerchief, with sleeves rolled up. Complete...... 25

55—THE MYSTERIOUS PERFUME BOTTLES AND HAND-KERCHIEFS, AS INTRODUCED BY BUATIER.

Fine sleight of hand work in vanishing and reproducing the handkerchiefs, using two borrowed hats and two perfume bottles. Very pretty and effective. You will like it.................. 50

56—THE HANDKERCHIEF PRODUCTION OR HANDKERCHIEFS FROM EMPTY HANDS.

Performer with sleeves rolled up shows both hands empty, rubs palms together and instantly produces a silk handkerchief, repeats the rubbing and another handkerchief appears; this is repeated a third time. In this excellent trick hands do not for a moment go near the body, but are at arm's length. Front and back of hands are both shown empty to audience. Handkerchiefs can also be vanished instead of produced if that is preferable. Performed nicely with very little practice 50

57—MARVELOUS PRODUCTION OF A HANDKERCHIEF.

One hand is all that is necessary to accomplish this trick, and your hand is kept in view of audience all the time; you show front and back to prove that you carry nothing. Your coat can be removed if desired. Now gradually produce a handkerchief, which you pass to audience for examination........ 75

58—THE SNAKE.

A very *charming* illusion. Performer takes a handkerchief which diminishes in hands, then entirely vanishes, and changes into a silken, wriggling snake.................................... 2 00

59—BUATIER'S FLYING HANDKERCHIEFS AND DECANTERS.

A very fine experiment with crystal decanters. *Performer* appears with two decanters, one in each hand. Decanters are inspected, also a silk handkerchief which performer thrusts right into one of the decanters and holds it free from the body. At command the handkerchief leaves one decanter and is found in the other. Complete............................ 3 50

60—SPIRITUAL DECANTERS.

A decanter is placed on your side table. You hold a second decanter in your hand in which you place a silk handkerchief, and going into midst of audience you command handkerchief to leave the decanter you hold and go, with imperceptible *rapidity, to decanter* on table, which it instantly does. This has a most marvelous effect, as decanters are in no way covered. Complete.................................... 3 50
Both of above together................................ 5 00

61—INVISIBLE FLIGHT OF HANDKERCHIEFS.
No. 1.

New and startling. A small square glass box, all sides transparent glass, is shown and covered with a borrowed handkerchief. Three colored handkerchiefs or flags are now produced in a new and novel manner; performer now makes a small cone of a piece of common paper or of one of the evening programmes, and visibly drops the handkerchiefs in same and jams them down with his wand or a borrowed cane, then hands cone to a spectator to hold. On uncovering the glass box the handkerchiefs are found in it, and are taken out and when cone is unrolled, the handkerchiefs are gone. Easy and a novelty...................................... 5 00
Ordinary quality...................................... 3 00

62—INVISIBLE FLIGHT OF HANDKERCHIEFS.
No. 2.

Same as No. 1, except that two small paper cones are used, being rolled up in full view of audience, and handkerchiefs dropped into one, disappear, and re-appear in the other, thus doing away with the glass box...................................... 1 00

63—HANDKERCHIEF AND CIGARETTES.

A new impromptu trick. Casually offer a cigarette from case, and taking one yourself, commence smoking. Now say, "Have you seen the latest vanish for a handkerchief?" A silk handkerchief is now vanished by an *entirely new method*, and found in the cigarette case, *cigarettes having disappeared.* The handkerchief is now shown for examination, after which it is gathered up in the hands and shaken out again when several cigarettes fall from its folds. Complete................ 2 00

Secret only, you can easily make it........................ 50

64—AN ABSORBING FEAT.

A small nickel plated box is shown around empty and placed on your table. One or two borrowed handkerchiefs disappear from the hands, and on opening box are found inside. New and creates great astonishment............................ 3 00

65—HANDY HANDKERCHIEF HANDLER.

An exceedingly useful invisible apparatus for vanishing or producing a handkerchief. Entirely new and first class. Unless specially ordered it is furnished for left hand work only... 1 50

66—A BRILLIANT PRODUCTION.

Artist shows hands empty and arms bare, yet produces from finger tips of left hand an immense quantity of ribbons. Entirely new and different from effects of this kind offered before. Ribbons of various size and colors can be used. Complete.. 2 50

67—"OLD GLORY," HANDKERCHIEF STAR.

Performer produces from between the fingers and thumb of left hand six long vari-colored rays, each one twelve to fifteen inches in length, forming a very pretty star, in the center of which suddenly appears a previously borrowed handkerchief or rings. New and excellent............................ 4 00

68—NEW HANDKERCHIEF WANDS.

With one of these wands any performer can at once produce one or two handkerchiefs or flags in empty hands. Very useful. 75

69—NEW VANISHING HANDKERCHIEF WAND.

This wand instantaneously vanishes any small handkerchief or flag without covering.................................... 75

70—HANDKERCHIEF WANDS AND HANDKER-CHIEFS.

Above tricks numbers 68 and 69 complete with silk handkerchiefs and dummy wand............ .... 2 50

71—NEW HANDKERCHIEF TO EGG.

This is a new invisible apparatus which changes any borrowed
handkerchief to a genuine egg; or used to vanish a real egg,
or handkerchief. A splendid piece of useful and perfect mechanism. Should be in every performer's hands. Entirely
new and first-class.. 2 50

72—COUNT PATRIZIO'S FLYING HANDKER-
CHIEFS.

A changing handkerchief feat entirely new to American Conjurers.
Two small handsome tubes are shown empty, each tube has a re-
movable top and bottom. A borrowed handkerchief is
in one tube, and a handkerchief of different color, also borrowed, placed in the other tube; on command the handkerchiefs change places instantly. Very striking effect, easily
executed. First time offered. Complete................... 3 00

73—THE EGYPTIAN TUBES. A MYSTERY.

A small package is suspended between two chairs by ribbons.
The artist borrows a lady's handkerchief, which vanishes from
his hands; hands shown perfectly empty. The package is
opened and found to contain a nest of six tubes, each tube
wrapped up and tied in paper. These tubes are opened one
after the other, the smallest one being opened by owner of
handkerchief who finds therein her own handkerchief. No
substitution of any kind. New and excellent effect.......... 5 00

74—NEW COLORING OF HANDKERCHIEFS.

Performer hands for examination his wand, a small leather tube
and three handkerchiefs. Then holding the tube so audience
can see through it, he inserts one of the handkerchiefs in tube
and pushes it through same, it comes out the other end still
white. He then holds tube so the audience can see through
it, and places the white handkerchief in the tube and pushes
it through, when it comes out at the other end red. He again
holds tube so audience can see through it, puts another handkerchief through in the same manner, when it comes out blue,
and the other handkerchief goes through the same operation.
The tube and wand are again handed for examination. First-
class and guaranteed superior to all changing handkerchief
tricks. Price... 3 50

75—FLAGS AND HANDKERCHIEFS.

American silk flags. Finest silk, warranted to wash and not lose
color.

Size			
7x10 inches, each..................................	15
" 8x12 " "	25
" 12x18 " "	40
" 16x24 " "	60
" 24x36 " "	1 00
" 32x48 " "	2 00

Special prices for a number and for larger sizes.
Cuban silk flags, size 12x18 inches, each.................... 40
Silk flags of all nations,
Size 12x18 inches, each.................................... 75
"   24x36   "   " 1 75
Finest of silk handkerchiefs, all colors,
12 inches square.. 35
16   "   "   45

76—THE WIZARD'S TRAVELING HANDKERCHIEFS.

Two glass cylinders are shown and handed to audience for examination; both are perfectly transparent. One mauve silk
and one red silk handkerchief are placed in one cylinder and
the lid tightly put on. A lady is requested to hold this. Performer now shows the other cylinder, which is exactly like the
first, only empty; lid is also put tightly on this and a gentleman is asked to hold it. He now borrows two handkerchiefs
with which he covers both cylinders, one over each, as they
are still held by lady and gentleman. He now commands the
two silk handkerchiefs to travel instantly from the cylinder
the lady is holding and into the empty one which gentleman
holds. Borrowed handkerchiefs are now removed from both
cylinders, and the one lady holds is found empty while the one
gentleman holds is seen to contain the two silk handkerchiefs.
Performer does not touch cylinders after covering them with
borrowed handkerchiefs. Cylinders can again be shown to
audience. This trick is novel and astonishing.............. 5 00
Included with above is the best handkerchief pull or vanisher, one
that can be used alone, and usually sold for................ 3 00

WITH WINE OR WATER.

"LET ME HAVE SUCH A BOWL."—*Henry VIII 1:4.*

77—THE PRISMATIC WATERS OF EBLANA.

This new and beautiful illusion seems specially adapted to con-
fuse the minds and deceive the eyes of the foremost pio-
neers of science. A tray of glasses is laid on side table.
Performer appears holding to full view of audience a transparent crystal decanter in which is pure water. One of
audience is requested to inspect the glasses, to prove that
they contain no chemicals, etc. Then performer pours into
the glasses different colored waters, and then asks a gentleman in company to do the same, but fails to produce anything but clear water. However, performer on receiving
back decanter, continues to pour other colored fluids from

same. The water may be drank to prove its genuineness, and decanter and glasses examined at any moment. Very easy and astounding... 1 00

78—INK AND WATER VASE AND CARAFE.

As performed by Hartz and Kellar. Vase of ink on table changes to water, and carafe with water, in hands of audience, changes to ink. A lightning change without covering vase. Complete... 5 00

79—THE FEAST OF BACCHUS.

Performer appears with any transparent crystal decanter filled with water, and inviting the company to partake of the "cup that cheers," requests them to make their own choice as to which wine they prefer—Hock, Moselle, champagne, port, sherry, or any other sort—it is all the same to performer, for he supplies their demands from decanter. The wines are poured out as called for and given to audience, who *con animoso* declare the genu-*wine*-ness of same. A first-class distribution illusion, and worthy of the *cordial*-ity it always receives. Complete without decanter, but with necessaries for several performances... 3 00

80—THE CHINESE RICE DISHES.

Two ordinary china dishes are used. One is filled with rice, and to prove there is no deception, the rice is poured out and the inside of the dish shown. The rice is then returned, and being over an inch above the top of the dish, is leveled off, making it just full; the second dish is then placed over it, but upon being instantly removed the rice is found to have doubled in quantity. Again it is leveled off and the same dish placed over it; this time when removed the rice has entirely vanished, and instead of it is a dish full of water or wine, which at once can be served to audience. This is a very fine illusion, and for apparent dexterity it stands pre-eminent. Including unexcelled patter................................... 3 00

81—FISH-BOWL PRODUCTION.

Bowls of water and fish from a handkerchief, each.............. 2 50

82—HOUDIN'S INEXHAUSTIBLE BOTTLE.

As introduced by Houdin. From one bottle performer pours four or six kinds of genuine wine or liquors. Small for four kinds 1 75
Large for six kinds...................................... 3 00

83—INK PILLS.

Valuable to all. When dropped in water it is instantly changed to ink. 100 pills... 3 00

WITH FLOWERS

"Then Shall I Raise Aloft the Milk White Rose, with whose Sweet Smell the Air Shall be Perfumed."—*Moore.*

84—FLOWERS FROM A PAPER CONE.

Buatier DeKolta's invention.

A.—Ordinary quality per hundred (100), single flowers...... 3 50
B.—Better quality per 100, double flowers.................. 5 00
C.—Finer quality per 100, " " 6 00
D.—Finest made, silk per 100 " " 8 00

85—BOUQUET ON PLATE.

Performer instantly produces on any plate a large, handsome bouquet of flowers without covering plate. New and striking, with flowers... 3 00

86—NEW FLAG AND FLOWER FEAT.

Performer produces from empty hands a good sized United States silk flag, rolls up his sleeves, and rubbing flag away between his hands it is seen to vanish entirely and change into a large, handsome bouquet of flowers. New and fine, for any performer. Complete with flag and flowers. Finest flowers 5 00
Second quality flowers..................................... 3 50

87—NEW FLOWER TRAY.

Small, handsome tray, nickel plated, produces flowers in or loads a hat with small articles, without covering. Two loads...... 4 00

88—APPEARING AND DISAPPEARING BOUQUET

Borrow a handkerchief from audience, and produce from same a large, exquisitely made bouquet, then vanish same suddenly from hands without being in any way covered. Bouquet looks perfectly natural. All made of feathers.............. 5 00

89—ORIENTAL FLOWER PRODUCTION.

Performer produces from any large handkerchief or small shawl a very large flower pot with magnificent growth of flowers, forming an elegant bouquet. Bears close inspection; done anywhere; the best production feat without apparatus extant; puts all others in the shade. Complete with flowers, each... 5 00
Pots alone to attach your own flowers to.................. 2 00
Two or three can be produced with one handkerchief.

WITH THE LADIES' FAVORITES.

"WELL DONE, MY BIRD."—*Tempest IV. I.*

90—THE NEW BIRD OF BACCHUS.

Performer appears with a *genuine glass bottle of wine* in hand, from which he pours out some of the cheering stimulant. He, however, previously borrows several rings, which are broken up and placed in a pistol, and performer asks one of audience to fire at bottle; but before doing so he pours out another glass of wine and gives it to the marksman to drink. Then the pistol is fired, and bottle broken with a hammer, when, to the delight of audience, a dove is found inside with the borrowed ring attached by a silken cord to its neck. This is a *glass* bottle. A very pleasing and wonderful feat. No assistant required..$ 1 00

91—THE CANARY'S FLIGHT.

Performer appears with a live canary, and turning up sleeves wraps canary in a piece of paper, then laying same on ground stamps it with his heel. On opening paper no bird is to be seen, but in its place feathers are discovered. A very simple and startling trick.................................... 1 50

92—JAPANESE BIRD VANISHING.

A new and strictly original manner of vanishing a bird or small dove. Can be done anywhere and at any time. Without exception the finest and most striking feat of its kind. Invention of the late renowned Prof. D'Alvini.................. 2 50

93—MEPHISTO'S BIRD CAGE.

Shown perfectly empty, at command it is filled with live birds.... 5 00

94—D'ALVINI'S JAPANESE BIRD CAGE.

A handsome brass bird cage is fully examined, and while empty is placed on a slender solid column and covered with borrowed handkerchief. On removing the handkerchief, one or two birds have appeared in the cage. Striking effect, as apparatus is all shown to audience and there is no apparent possibility of concealing a bird in it. First-class effect...... 5 00

95—THE VANISHING DOVE CAGE.

A handsome nickel plated cage, of size to contain a dove, is handed to one of the audience to put your dove or several canaries in, and on its being returned the performer places a handkerchief over it, and swaying to and fro, it immediately disappears. Finest made and a great favorite............ 5 00

A pair of above cages with an exceedingly novel and laughable manner of reproducing the one just vanished. Price........ 8 00

96—BUATIER'S FLYING CAGE.

This celebrated trick, which still creates most profound amazement, is the favorite of many professionals. The cage is of brass or nickel-plated wire, and contains a live bird. Instantly while the audience is staring at it in hand of performer, it vanishes without any covering whatsoever, leaving no trace behind. Many of the cages which are sold are very clumsily made. Every cage purchased of us is guaranteed in working order.. 5 00

Same, extra fine... 8 00

97—THE MYSTIC PRODUCING CAGE.

This handsome cage is finely made in polished wood and brass wires, every part of it shown to audience, yet produces in it a number of canary birds or a pair of doves, works either in the hand or hanging up, no covering required. Very effective and useful. Price..................................... 5 00

98—THE DOVE OR PIGEON PAN.

Regulation Professional stew pan, nickel-plated, for two or three doves,—the finest made................................. 5 00

99—THE ENCHANTED BIRD BOX.

Changes eggs to birds, or produces, changes,or vanishes any small articles. Top and bottom open, inlaid, a fine piece of apparatus, useful in combinations............................. 8 50

100—THE FEATHERED MESSENGER.

Performer rolls up a small cone of piece of newspaper, and suspends it from a hook attached to a silk cord. He then shows several narrow strips of tissue paper, which he rubs between his fingers, when they immediately change to feathers, these are rubbed a little more and they change into a canary bird, but, alas! it's dead! Artist now loads it into his pistol and shoots at the paper cone, out of which the live canary instantly flies.

New and excellent, complete except live canary and pistol. 5 00

With live canary and pistol............................. 9 00

101—THE BIRD CASKET.

A crystal casket, all sides glass, containing live birds, is placed on a table. An empty cage is then hung at some distance and covered with a handkerchief. In an instant the birds disappear from casket and are found in the cage. 12 00

102—THE MIKADO'S FAVORITE.

The latest sensational trick from the Land of Marvels. Should be in the hands of every professional. Two small elegant pedestals are shown empty; audience can examine them; on one is placed a beautiful cage with bird and covered with a borrowed handkerchief, and handed to a spectator to hold; on the other empty pedestal is placed a borrowed handkerchief, and this also handed to a spectator to hold; now a wonderful change takes place, for as soon as the spectators remove the handkerchiefs themselves, the cage and bird have completely vanished from where they were placed, and are found on the other pedestal, from which cage is removed and handed to audience for examination to show it is solid.................. 15 00

102 A—THE LONDON PHOENIX.

As *originally* introduced in Egyptian Hall, London, by Mr. David Devant. Artist shows a cage containing a live canary and a small paper bag; he removes bird from cage and places it in this bag, then places bird and bag on a target which is suspended in mid-air. After showing that the target is quite isolated, he brings forward a gun, and offers it to anyone who will undertake to hit the bullseye. A gentleman fires accordingly, but evidently does not hit the bulls-eye, for on the contrary he puts out the light of a candle standing on a side table. The artist takes a shot at the candle, which immediately lights it again. Once more the gentleman tries his luck, and with a startling result, for the paper bag containing the bird bursts into fire, and the target instantly changes into a large bird-cage with the living bird contentedly hopping about. Splendid effect. Complete, square cage 15 50
Round cage.. 23 50

102 B—THE LATEST IMPROVED PHOENIX.

A small cage containing a canary has been standing on the table from the commencement of the performance. When about to present the illusion, performer hangs up the target by two hooks. He then wraps the small cage *containing live canary* in tissue paper, and places it on a little stand on top of target. He then sets fire to the tissue paper, a slight explosion follows, the target *disappears entirely*, being replaced by a large cage, which contains the live canary, and the little cage is empty. No assistance required, nothing to get out of order. Can be done in a parlor. A fine portable illusion, complete and elegant. Price and further particulars on application.

103—BIRTH, DEATH AND RESUSCITATION.

Performer borrows a handkerchief, places it on end of wand when it suddenly vanishes in a flash of flame. He then produces one or two eggs, rolls them up in a piece of tissue paper, breaks eggs, paper has changed to the handerchief, and a canary appears, which is placed in a small wire cage. A large silver or nickel-plated vase is shown empty, and filled with bran; on shooting at cage the bird falls dead, it is then taken out and buried in the bran in the vase, from which there now slowly grows a large, magnificent bouquet, and from center of it rises a tulip which slowly unfolds and out of it flies the bird. There are several different combinations or variations of this fine trick. Price according to effect desired from $25.00 to..$ 50 00

WITH A BORROWED HAT.

"By My So Potent Art."—Tempest V. I.

104—ARTICLES FROM A BORROWED HAT.

Mantel clocks......................................each	$	1	00
Six champagne bottles..................................		3	00
Three glass lanterns...................................		10	00
Life size baby..		4	00
Large growth of flowers...............................		5	00
Six glasses...		1	50
Hollow skull, used by M. Hartz and recommended by Professor Hoffmann......		5	00
Genuine human skull...................................		10	00
Solid cannon ball.....................................			75
Hollow cabbage.......................................		2	50
Tin goblets, per dozen................................		1	25
Brass goblets, per dozen..............................		1	75

105—MECHANICAL HAT LOADER.

Can be attached to any table and loads hat or cone without performer going behind table. Indispensable............... 2 50

106—BEST MAKE RAREBIT SAUCEPAN.

A great favorite.. 2 50

107—THE FLOWER BAKERY.

"Now we drop eggs, flour and all necessary condignments—I beg pardon, condiments—into this gentleman's empty hat, hold it over this column of grease (a lighted candle) a few seconds, tip contents onto this plate, and behold, a magnificent bouquet, with the ladies' rings attached." Excellent............ 3 50

108—SWEET WILLIAM'S HAT.

On a glass stand are placed some flower seeds. The stand is covered with a borrowed hat, and, on lifting same, instantly a large, handsome bouquet has appeared on the stand much larger than hat itself. All complete........................ 5 00
Per Pair.. 8 00

109—A LIVELY SUPPER.

From our own programme. An entirely new and entertaining way of introducing the amusing trick of "Cooking in a Hat." First-class for stage or parlor, with patter, apparatus, cages, doves, all complete. An excellent combination............ 20 00

THOSE MYSTIC BOXES.

"FOR WHAT I WILL, I WILL"—*Two Gentlemen of Verona, I. 3.*

110—DISSECTING DICE BOX.

This is one that puzzles the smart fellows who "know it all," and it should be in the possession of every professional. The old die trick is first shown, and then the solid die placed in the dissecting box, which has two compartments, with two doors to each. The die appears in either one of them, yet vanishes completely at will of performer, and box shown completely empty. Elegant finish only. Price........................ 8 00

111—JAPANESE INEXHAUSTIBLE BOX.

Every performer appreciates the value of this. It is elegantly made and finished. Shown empty and from it is immediately produced any quantity of articles, and is again shown perfectly empty, yet still another supply is produced from it, it being as its name indicates, *"Inexhaustible."* Price.............. 7 00

112—DISSECTING AND INEXHAUSTIBLE JAPANESE BOX.

An invaluable article for all performers. The box consists only of four sides and bottom, no cover. It is put together in view of audience. From it, is at once taken or produced any quantity of ribbons, handkerchiefs, balls, etc., etc. It is then taken to pieces and again put together, when from it is produced cages, birds, lanterns, doves or rabbits. No artist should be without it. Price of box alone.................. 5 00
Extra large size, per original model of the inventor Prof. D'Alvini, in our possession.................................. 8 00

113—THE NEST OF BOXES.

A beautifully-finished box is brought forward by the enchanter, who locks same and gives key to one of company to hold. A watch and handkerchief are then borrowed, placed in a pistol, and pistol discharged at the box. Upon unlocking the box and opening it there are found six other boxes, each of them locked, in the last and smallest of which are found the watch and handkerchief. Very useful for many tricks. Price $10 00, $15 00 and 25 00

114—THE CRYSTAL BALL-BOX AND VASE.

An elegant crystal casket, the lid and sides of which are all glass, so that it is apparently impossible to conceal anything in them, is shown to be perfectly empty. It is suspended by two light cords in mid-air. A glass vase is then shown full of vari-colored balls. At command the balls disappear from vase, and in a most mysterious manner they appear in crystal casket. This is a great favorite with magicians, and a most astonishing trick. Complete, extra fine finish.............. 25 00

115—TEMPUS FUGIT, OR "TIME IN A FIX."

A new and most excellent trick. First is introduced to audience a handsome box, locked, corded and sealed and held by one of the audience. A borrowed watch is wrapped in a hand-kerchief, after first being rolled up in *any* color of paper and tied with *any* color of ribbon as selected by audience, then handed to *any* person to hold, and he not only feels the watch in the handkerchief but *hears it* ticking also. On shooting at box the handkerchief is found empty, and on box being opened is found to contain another, which contains a third one in which is found by the audience the same watch wrapped up in same paper and tied up with same ribbons as before. No confeder-ates, no table work required. Boxes finest finish and bear minute examination. So called explanations of this have been "going the rounds," but *not one* is correct.............. 15 00

116—THE CRYSTAL CASKET AND BALL BOX.

An elegant crystal casket, the lid and sides of which are all glass, so that it is apparently impossible to conceal anything in them, shown to be perfectly empty. It is suspended by two light cords in mid air. A beautiful box is then shown full of varicolored balls. At command the balls disappear from box, and in a most mysterious manner they appear in crystal cas-ket. This is a great favorite with magicians, and a most as-tonishing trick. Extra fine finish. Complete.............. 25 00

117—THE CRYSTAL CASKET AND CANISTER.

Similar to preceding trick, except instead of ball box a hand-some large canister is used; this is first shown empty, filled with balls, which disappear without covering, and appear in the crystal casket. From the empty canister is then produced two or three large cages containing live birds. Price, according to capacity required, $20 00 to.................... 25 00

Both of above tricks have been used by many prominent performers, including Professors Seeman, Reno, Bancroft and others.

MISCELLANEOUS.

"IF THIS BE MAGIC, LET IT BE AN ART."—*Winter's Tale. V. III.*

118—PROGRAMME DESIGNS.

Handsome copy-righted designs for house programmes or handbills. Two sizes. Plates furnished. Samples free for stamp.

119—THE MAGICIAN'S ENVELOPE.

A fine double envelope, yet in appearance a single one. Useful for producing, vanishing or changing articles............... 10

Same in newspaper form for larger articles.................... 10

120—APPARATUS FOR NEW FLOATING WAND.

Fine and cheap... 35

121—PREPARED CANDLES.

Prepared hollow imitation candles from which cards, handkerchiefs, cigars, etc. can be produced. Very useful and attractive. Per dozen..................................... 60

122—A MAGICAL SHOT IN REALITY.

To instantaneously catch a handkerchief and an egg in your mouth, when shot from a pistol; something entirely new and very amusing.. 25

123—REAL EGG FOR VANISHING HANDKERCHIEF OR GLOVE.

Very useful... 25

124—THE MYSTIC KEY.

One of the most interesting tricks. One or more finger rings can be placed upon the key and removed without injury to either, although beard of key is larger than the rings. Brass, finely finished... 50

125—NEW SHOOTING WAND.

Takes place of pistol, can be examined before firing. Indispensable for creating fine effect.............................. 1 50

126—NEW TRAP TRAY.

Small brass tray, nickel plated; very innocent appearance, yet contains a trap to vanish articles through into a hat or other receptacle without covering; useful...................... 3 00

127—TURTLE DOVES.

Nice and small, accustomed to handling. Bred in our own dove cote. Per pair... 3 00

128—GUINEA PIGS, PURE WHITE.

Pink eyes, superior to rabbits for many reasons. Our own breeding. Per pair... 3 00

129—LONG HAIRED PERUVIAN GUINEA PIGS.

Pure white, pink eyes, superior to rabbits for many reasons. Our own breeding. Per pair.............................. 4 00

130—THE FIERY FLASHING RINGS.

Several borrowed rings are wrapped in a small piece of paper and given to a spectator to hold, and while they are still in his hands, at command of performer they vanish instantly in a flash of fire, leaving absolutely nothing in the person's hands, to his intense astonishment. Rings can be produced elsewhere as the ingenuity of the performer may suggest. Price for full manner of working and necessaries............ 2 00

131—NEW CHANGING RING TRAY.

See number 51. But for rings instead of coins................... 3 50

132—CHINESE LINKING RINGS,

Fine steel, nickel plated; diameter of rings 9 inches, thickness about ⅜ inch. Eight rings in the set; link in a most mysterious manner. *The best rings made.* Per set............... 6 00
Smaller set... 2 50

133—THE PHANTOM ORANGE.

Performer hands for examination two brass cylinders open at each end, also an orange; he next shows two common china plates. Placing plates upon table, he holds cylinders so audience can see through them. He places one cylinder on one of the plates, and slips orange into top of the same, then lifts cylinder so that audience can see orange on plate; he then places the other cylinder on the other plate, and upon lifting the cylinder over the orange, it has disappeared, and lifting the other cylinder, the orange is seen on the other plate. The plates and cylinders are not prepared in any manner. No covers are used. Orange changes position at will.......... 3 50

134—AN INTERCHANGEABLE TRICK.

"Nous avons change tout cela."

Performer rolls a sheet of newspaper into a cone and drops a handkerchief into it. From his wand he produces a billiard ball, which he vanishes and it appears in the cone instead of the handkerchief. Replaces the ball in the cone, then takes a handkerchief in his hands which changes to a billiard ball, and handkerchief is found in the cone from which ball has vanished.

Entirely new and of first-class effect. Complete........... 5 00

135—A FUNNY CANDLE.

Performer lays a lighted candle on table. Suddenly the light goes out by accident, I suppose, but wishing for light, the magician blows at it, when it instantly relights and as rapidly is extinguished. He blows at it again, when it sheds forth its luminous rays. This may be repeated any number of times..... 3 50

136—THE CHARMED PISTOL TUBE.

Is superior to any used, because the audience sees borrowed articles really put in tube; they are not removed in any way, and still when pistol is fired tube is shown to be perfectly empty, and articles must therefore have been fired out of it. Made in brass. Effect is wonderful!............................. 5 00

137—PRODUCING TRAY.

Small ebonized tray, to produce cards, envelopes, answers or handkerchief, after being shown empty and without covering. 7 00

138—NEW TRAY FOR CHANGING WATCHES.

See number 51, but for watches instead of coins................. 10 00

139—COMBINED CHANGING COLLECTORS.

For changing watches, rings and handkerchiefs, all at once...... 10 00

140—NEW INEXHAUSTIBLE SARATOGA TRUNK.

Specially designed and used in the Dramatic Order Knights of Khorassan, K. P., by our Mr. H. J. Burlingame, Master of Mysteries of Al Hathim Temple, No. 24. An entirely new method of introducing an improvement on the old Inexhaustible Japanese Box, making it, in fact, a new illusion. Trunk is handsomely made, imitating the familiar Saratoga style. Carries sufficient articles for a fifteen or twenty minutes performance, and produces birds, cages, flags, handkerchiefs, fruit, lanterns, bottles of liquors, flower-pots, flowers, cigars; in fact, almost everything imaginable, though shown empty after each production. Especially adapted for ladies or for short turn on a programme, being a complete show in itself.

Requires no skill. Is being produced with marked success before critical audiences. Price of trunk alone............... 10 00
Price of trunk with complete outfit from ....20 00 to 50 00

141—THE HEAD OF IBYKUS.

Two chairs are placed back to back, about twelve inches apart; on tops of backs is placed a sheet of glass, on which latter is a skeleton head. This head answers any questions propounded by either a nod or a shake. A glass dome is now placed over the skull, still it continues in its peculiar antics. It deliberately turns around and stares at some one in the audience, and reveals very startling stories relating to the person it is looking at. It smokes, whistles, opens its mouth, chatters its teeth and in short, like all other "dead heads," knows too much. It is an exquisite model of superior finish, and is not, as many fancy, worked by electricity. Complete.. 20 00
Same as above, head only.................................... 3 50
Same as above, genuine skull................................ 10 00

142—NEW SPIRIT DIAL.

Polished plate glass clock dial, with pointer; gold numbers on dial which is suspended by cords or on pedestal. Pointer or hand revolves and stops at any number desired. Can be taken off, replaced and revolved by a spectator, yet it will stop at any number desired. Extraordinary useful and puzzles any audience. Our dials are in constant use; none better made. Dial alone, to work hanging up or held in hands, $8.00 and $15.00. Same with handsome gilt pedestal $20.00 and.............. 25.00

143—SPIRIT DIAL AND TALKING BELL.

Above dial, combined with Crystal Talking Bell, each working separately, $20.00 and.................................... 25.00
Both together in one elegant, portable, showy, polished brass frame 25 00

144—WANDS, TRAPS, TABLES AND APPLIANCES.

Rubber Oranges, each...................................... 25
Real Hollow Eggs, reinforced, each......................... 25
Rubber Eggs, each... 50
Wands, plain or ornamented 25 c to........................ 1 50
Plain trap, $1.50 and...................................... 2 50
Wrist   " 4 00
Rabbit trap single $3.50, double........................... 6 00
Tables, single leg, cast feet, $8.00 to..................... 15 00
Our best side table, new cable design, fancy brass tubing, the finest table made, without exception, goes in your trunk. Each $20 00 and... 25 00
New cable design center table; black and gold, or all gold; magnificent ornaments and fittings; the most original design ever

conceived; complete in every respect; also forms two superb side tables when desired; are unequaled; with case 75 00
Louis XIV tables, any desired finish 75 00
New loading tables, to produce any ordinary article, such as doves, balls, bowls, flowers, pots, fowls, tea or coffee service, or anything from a hat or handkerchief, light and portable; never offered before; very useful and first class. Has neither shelves nor traps, a fine novelty 12 00
Photographs of tables on application with stamp. .
Lifting Trap Tables, for Nest of Boxes or Hat Loading, with shelf, top alone without drapery 5 00
Same complete 15 00
Tables (without shelf) to load hats or anything else, when performer is surrounded by audience, or in a circus ring 15 00

144 A—THOSE MYSTIC BALLS.

"As swift in motion as a ball."—*Romeo and Juliet, II, 5.*

B. Handkerchief changes to Billiard Ball. Very useful   75
C. The Disappearing Billiard Ball in Glass of Water 2 00
D. Improved Multiplying Billiard Balls 1 50
E. Improved Multiplying, Diminishing and Vanishing Billiard Balls 5 00

PSYCHICAL PHENOMENA AND ANTI-SPIRITUALISTIC ILLUSIONS.

"Secrets Now Veiled, to Bring to Light."—*Goethe's Faust.*

145—CLAIRVOYANT BOOK AND SLATE MYSTERY.

Of excellent effect, any book used, and anything selected in it by any person of audience, is found written on the slate previously shown bare. Far superior to that where any faked-book is used . . . . $ 1 00

146—THE NEW SPIRIT HAND FOR RAPPING.

This is an excellent model of a lady's hand and forearm, with lace sleeve and cuff. Bears minute examination, and as soon as placed on a table or sheet of plate glass, begins at once to rap out any numbers or answers desired. Excellent. Superior to any made 1 50

147—THE MAGI'S WAND.

An elegantly made wand, nickel plated, large fancy ornaments

on each end, can be examined, yet floats in the air without being touched, and handed to audience at any time. Something new and very effective............................... 2 50
Secret.. 25

148—TABLE AND CHAIR LIFTING UNEXCELLED.

By placing your hand in centre of top of table, chair, stand, box or barrel, you can instantly lift up same and carry it around clinging to palm of hand and allow audience to remove it. Done anywhere, light or dark, with sleeves rolled up. Excellent. Price. including floating hat........................ 3 00
Cheaper system... 1 00

149—NEW SPIRIT SLATE WRITING.

This is a first-class slate writing feat for *close* circles. Any ordinary slates are used and answers to questions asked by investigators appear on one of the slates. In *small circles* this is one of the best slate writing feats extant. Unknown to professionals.. 1 50

150—THE ARABIAN NIGHTS DIVINATION.

That wonderful volume of mystical lore known to even the smallest heirs as the "Arabian Nights," is produced by performer, and given to audience for inspection. One of company is requested to select any page, or number to indicate page, and another is invited to choose any line or number to indicate line. Performer instantly hands the volume over to some lady or gentleman present, and requests that she or he open page at desired number and to see the chosen line.

When this is done performer at once without any confederacy tells the page, the line, and even recites it. A most marvelous deception.. 2 00

Secret.. . 50

151—THE MARVELOUS BOOK MYSTERY.

Unquestionably the best and most original book mystery ever introduced. WITH SPLENDID PATTER. First, a large blank slip of paper is put in an envelope, sealed and handed to one of audience to hold. From several volumes of prose or poetry, one is selected at random by any person, and any verse or page selected by any one of audience, is at once read aloud by your lady or medium, who is on stage blindfolded, and while book is in hands of audience, and envelope opened and the blank is found to contain the entire verse or page written out, or even a still more sensational ending can be used. Excellent. No prepared books and no confederates.. 2 50

152—THE LATEST AND BEST SLATE WRITING.

With this you can write on *any* slate at any time and place, and even while the end of slate is being held by investigator. There are no flaps, no chemicals, no ringing the changes of slate and no leaving the room in this; it is simple and easily executed. Can be used on stage, two methods, and is without exception the best thing in this line ever brought out. Price.......... 5 00

153—THE FAMOUS POST TYING SEANCE.

Performer is securely bound, sewed and sealed by any cord, rope, thread or cloth to a bolt solidly screwed up in a post by audience. Excellent. Price.................................. 9 00

154—NEW THOUGHT READING PADS.

Entirely new Thought Reading Pads, best made, bear close examination, yet reveal anything written, including Anna Eva Fay's system. Per 100.................................. 10 00

155—NEW SPIRIT BOOK READING.

Medium blindfolded on stage. A book of poems or prose is handed to any person in audience to examine, another person denotes at what page or place the book shall be opened, which is done, and the medium at once recites the page, chapter, verse, lines, etc. Not to be confounded with similar effects previously offered, this being different from all of them and easily executed. Splendid effect. Complete with book (selection of classics allowed).......................... 6 00

THE SPIRITS AT WORK, Sensational Spiritualistic Seance.
"THE BEWITCHED VALET, OR A NIGHT WITH SPIRITS."
A Fine Magical Act.
"A MEPHISTOPHELIAN CARNIVAL, OR A TILT WITH IN-
VISIBLE SPIRITS."
KELLAR'S CABINET MYSTERIES.
ANNA EVA FAY'S MANIFESTATIONS.
MATERIALIZATION AND DE-MATERIALIZATION.
SPIRIT COLLARS, BOLTS, STAPLES, LOCKS, HINGES, HAND-
CUFFS, POSTS, CROSSES, PILLORIES AND BENCHES IN
GREAT VARIETY.
PRICES AND DETAILS ON APPLICATION.

157—SPIRITS AT HOME.

Complete outfit for a first-class spiritualistic seance, for halls, par-
lors, small theaters, etc., including all the most phenomenal
feats, such as table lifting, rope tying, slate writing, vest tests,
new mind reading, spirit bolt, spirit rapping and writing hand,
sealed letter and book reading, and *many* others, including
all necessary apparatus without canopy, but including plan
of same if desired. The best selection a purchaser can make,
as it is a complete entertainment........................... 25 00

158—SPIRIT NET AND EASEL.

A large, handsome easel is shown and placed in center of stage; a lady reclines against this easel; a large net is placed over her and securely fastened to the easel, completely enveloping the lady. This is done by a committee, and the closest inspection allowed before and after. A screen is placed in front of lady, when all the usual spiritualistic manifestations take place, and she is finally found outside of the net, without the net or easel being disturbed. Done anywhere. First class sensation. Price...................................... 75 00

A FEW NEW ILLUSIONS.

"STILL FOR THE FOND ILLUSION YEARNS MY SOUL."—*Goethe's Faust.*

159—ELECTRIC DECAPITATION OR ELECTROCUTION.

This is an entirely new decapitation act, and can be done anywhere, in any hall or on any stage. No previous preparation necessary. No glasses, bulky apparatus or large upholstered chairs. A cane-seat chair used. Head taken completely off and shown around. Light and portable. Price, complete with electrical connections, head and packing case, $75.00 and 100 00

160—"QUEEN OF KOR," OR CREMATION.

A handsomely finished and decorated casket is placed on stage on a board on two trestles, and a committee from audience are invited to examine it thoroughly, inside and out. It has glass openings in front, so interior is seen when cover is shut. The performer's assistant (lady is preferable) steps up into it and cover is closed. Three swords are passed entirely through casket in three different directions, screams are heard, etc., etc., etc., lid again opened, and the whole interior of casket set on fire, which is allowed to burn down, then ashes are taken from it. On reopening the cover the lady is found to have vanished. A large rug is then brought in and laid on floor of stage, the casket lifted off trestles (and board) and placed on rug and examined by the member of audience on stage. He gets in and the professor digs sword in his body; he is then let out and casket placed by professor and assistant on trestles again. Professor now fires pistol, casket opens and lady steps out. Done on any stage or in any hall; no traps or glasses. Further particulars and price on application. General so-called explanations of this illusion have been floating around, *not one* of which is correct. This illusion has been and is being successfully used now, yet is new here.

161 — THE AERIAL FLIGHT.

Four tall uprights, nicely plated, are fastened on the stage; up and down in these uprights there moves a large handsome metal cage. A lady is placed in this cage, small curtains are hung around cage only, which is then raised to the top of the uprights. On shooting at the cage, curtains fall to the floor, and the cage is empty, lady having disappeared. *At no time does the cage or curtains come near the floor.* Can be worked on any stage. No traps or glasses. First-class sensation. Price $225 00

162 — THE NAVAL ECLIPSE.

A large cannon is placed on the stage, also a handsome pagoda. A young lady dressed in pink costume is loaded into the cannon, and a young lady dressed in blue costume is placed in the pagoda, which is now raised up in the air. Cannon is fired, the pagoda opens and discloses the lady in pink, the lady in blue having disappeared to reappear whenever you like. If preferred only *one* lady can be employed and this same *one* shot from the cannon *into the pagoda while it is suspended in the air.* This is first-class and sensational in every respect. Price 250 00

163 — LIST OF SPECIALTIES FOR STAGE.

THE FAMOUS DOUBLE INDIAN MAIL.
SELF-RISING REVOLVING AERIAL SUSPENSION.
THE BUNGALOW.
MADAME SANS GENE.
TRILBY, OR LEVITATION MYSTERY.
SHEIK'S PAVILION, OR MYSTERY OF THE DESERT.
HINDOO FAKIR'S TRANSMIGRATION.
LIVING CARD TARGET.
BIRTH OF FLORA.
TALKING BELL. WHY NOT? IT HAS A TONGUE.
MYTHIA. A CHARMING ILLUSION.
THE CAGE OF ENCHANTMENT.
THE LATE PROF. ANDERSON'S CRUCIFIXION ACT.
AND OTHERS.

164 — PROGRAMMES, OUTFITS AND SPECIAL ACTS.

We make a specialty of furnishing complete programmes with apparatus, tricks and patter for home circles, societies, churches or theatres; or short acts for vaudeville, also scenic outfits of any size whatever. Let us know what you desire and we will make estimates.

165 — LECTURES.

"Among the Dykes and Ditches of Holland," or Rambles over the Flower Garden of Europe. Without views, $20 00. With 50 Lantern Views.. 50 00

"On Foot through the Valleys and over the Glaciers of Switzerland," or Enchanting Pictures of the Marvelous Mecca of European Tourists. Without Views, $20 00. With 50 Lantern Views.. 50 00

"A Rap at the Rappers," or Modern Science vs. Spiritualism. The combined work of Sid Macaire of Dublin, and H.J. Burlingame and Col. Wm. Lightfoot Visscher of the Press Club of Chicago. Lecture alone $25 00. Complete with paraphernalia, to duplicate all the leading effects of the so-called Spirit-Media.. 50 00

ADDENDA.

THE ENCHANTED PRISON CELL,
or "The Haunted Page."

"SOMETHING WRONG,"
A Magical Interlude.

Above two sketches are among the best ever offered for a magical entertainment. They require from three to five persons each, and take up from half to three-quarters of an hour, the combined work of Mr Sid Macaire of Dublin, and Colonel William Lightfoot Visscher and Mr. H. J. Burlingame, of the Press Club of Chicago, and the Fakir of Burmah. Outfit for each complete, $250.00, including play. Play alone without apparatus, $25.00.

We have a few copies left of the paper pamphlet, "History of Magic and Magicians," by H. J. Burlingame. Price, ten cents each.